This is the first book to present a coherent theoretical and experimental treatment of the rapidly developing field of macroscopic quantum tunneling of the magnetic moment.

The theory is based on the concept of the magnetic instanton and its renormalization by the dissipative environment, and the book includes discussions of the tunneling of magnetic moments in small ferromagnetic grains, tunneling of the Néel vector in antiferromagnetic grains, quantum nucleation of magnetic domains, and quantum depinning of domain walls. The experimental part collects the majority of recent data that are, or may be, relevant to spin tunneling. Among the topics described are low-temperature magnetic relaxation and its interpretation in various systems, experiments on single particles and mesoscopic wires, and resonant spin tunneling in molecular magnets.

This study of an important new field in condensed matter physics by two leading contributors to the subject will be of interest to theorists and experimentalists working in magnetism, and will provide a sufficient background to allow either to begin independent research.

Cambridge Studies in Magnetism

EDITED BY

David Edwards
Department of Mathematics, Imperial College of Science, Technology and Medicine

Macroscopic Quantum Tunneling
of the Magnetic Moment

Macroscopic Quantum Tunneling of the Magnetic Moment

Eugene M. Chudnovsky
City University of New York

Javier Tejada
University of Barcelona

CAMBRIDGE
UNIVERSITY PRESS

CAMBRIDGE UNIVERSITY PRESS
Cambridge, New York, Melbourne, Madrid, Cape Town, Singapore, São Paulo

Cambridge University Press
The Edinburgh Building, Cambridge CB2 2RU, UK

Published in the United States of America by Cambridge University Press, New York

www.cambridge.org
Information on this title: www.cambridge.org/9780521474047

First published 1998
This digitally printed first paperback version 2005

A catalogue record for this publication is available from the British Library

Library of Congress Cataloguing in Publication data

Chudnovsky, Eugene M., 1948-
    Macroscopic quantum tunneling of the magnetic moment / Eugene M.
Chudnovsky and Javier Tejada.
        p.   cm.
    "February 11, 1997."
    ISBN 0 521 47404 3 (hc)
    1. Tunneling (Physics)   2. Magnetization   3. Instantons.
I. Tejada, Javier.   II. Title.
QC176.8.T8C48 1998
530.4′16–dc21 97-26060 CIP

ISBN-13 978-0-521-47404-7 hardback
ISBN-10 0-521-47404-3 hardback

ISBN-13 978-0-521-02261-3 paperback
ISBN-10 0-521-02261-4 paperback

To our parents,
Michael and Sofia Chudnovsky,
Raúl Tejada and Conchita Palacios

Contents

Preface

The book consists of two parts. The first part (Chapters 2–4) is purely theoretical; the second part (Chapters 5–7) is mostly experimental. Chapter 2 contains the general theory of instantons and tunneling with dissipation, which is necessary for understanding the rest of the book. Chapters 3 and 4 deal with magnetic tunneling in single-domain particles and bulk materials, respectively. In Chapter 5, the consequences of tunneling for magnetic relaxation are derived and applied to experiments. Non-relaxation experiments are discussed in Chapter 6. Data on resonant spin tunneling in $Mn_{12}Ac$, and their interpretation, are presented in Chapter 7. In selecting material for this book we were guided by the principle that the theory must be relevant to experiment while experiments must be relevant to magnetic tunneling. Time may prove that some are not. Lengthy theoretical formulas which are difficult to compare with experiments and experimental data the analysis of which is too complicated have been left out. No doubt, there are important results that did not enter the book because we failed to appreciate their significance. Our list of references serves the single purpose of refering the reader to the original papers that we know and understand; in no way does it constitute a complete list of important works on magnetic tunneling.

The work on this book would have been difficult or impossible without support from the National Science Foundation of the USA, the Spanish and Catalan governments, and the Banco Bilbao Vizcaya. We are also infinitely grateful to Joan Manel Hernandez for linking together a vast number of computer files during the final stage of the work on the book.

Chapter 1

Introduction

Traditionally, the physics of magnetism has been divided into two almost independent branches. The first branch deals with interactions at the atomic level.
When applied to the three-dimensional world, it mostly consists of challenging
unsolved problems like the Ising, Habbard, and Heisenberg Hamiltonians,
Fermi-liquid models of itinerant magnetism, etc. It is concerned with the
ground state and excitations relevant to small scales, typically a few tens of
atomic lattice spacings. The Heisenberg ferromagnetic exchange, for example,
leads to the formation of a constant local spin density and the quadratic dependence of the energy of short-wavelength spin excitations on the momentum.
This has certain consequences for the saturation magnetization, its temperature
dependence, the magnetic contribution to the specific heat, neutron diffraction,
etc. However, when one turns to the properties of a magnet on a mesoscopic
scale of 100Å and greater, most of what can be derived or predicted from the
microscopic theory becomes irrelevant. For instance, the practical question of
how a piece of iron magnetizes in an applied magnetic field has nothing to do
with the Heisenberg Hamiltonian. This is because processes observed on the
mesoscopic and macroscopic scales result from weak interactions unaccounted
for in simple quantum-mechanical models. They are the magnetic anisotropy
owing to the symmetry of the crystal, the magnetic dipole interaction that
breaks the magnet into magnetic domains, interactions of domain walls with
defects, impurities, itinerant electrons, etc. At that level, the only relevant prediction of a microscopic theory is that a solid possesses a non-zero local spin
density, that is, local magnetization, or any other spin order parameter, e.g.
the Néel vector in antiferromagnets. The branch of magnetism that studies
mesoscopic scales is called micromagnetism. Its fundamental equation is the

equation of motion for the magnetization, $M(r, t)$, proposed by Landau and Lifshitz in 1935 [1]:

$$\frac{\partial M}{\partial t} = -\gamma M \times \frac{\delta E}{\delta M}.$$ (1.1)

Here E is the classical energy of the magnet and γ is the gyromagnetic ratio. The generalization to ferrimagnets and antiferromagnets is obtained by introducing two or more ferromagnetic sublattices, M_1, M_2, etc., each satisfying Eq. (1.1), and the energy of the magnet, E, that depends on the magnetizations of the sublattices and their spatial derivatives. In this approach, the form of the energy is derived from the symmetry arguments, with phenomenological coefficients determined by experiment. This purely classical method has been amazingly successful in describing magnetic phenomena down to scales as small as a few atomic lattice spacings. It concerns equilibrium magnetic configurations, the motion of domain walls, ferromagnetic resonance, spin waves, etc. The power of the method has its origin in the strong exchange interaction between individual spins of a typical magnet. Owing to that interaction the total magnetic moment of a few thousand atoms behaves as one entity. Still, it satisfies the quantum commutation relation,

$$M_i M_j - M_j M_i = 2i\mu_B \epsilon_{ijk} M_k,$$ (1.2)

but as soon as M significantly exceeds the Bohr magneton μ_B the right-hand side of Eq. (1.2) becomes small compared with each of the two terms on the left-hand side of this equation. The greater M compared with μ_B the better the classical approximation that assumes that the components of M commute with each other. For a typical ferromagnet, the magnetic moment of a nanometer-size grain is already a few thousand Bohr magnetons. It can then be considered classical in the sense that the orientation of M (that is, all three of its components) can be determined with high accuracy by a macroscopic measurement. In that sense a nanometer-size ferromagnetic particle is qualitatively similar to the arrow of a compass.

The question addressed in this book is that of the circumstances under which one can observe quantum tunneling of a large magnetic moment between classically stable magnetic configurations. There are several reasons why this problem deserves attention. The most trivial reason is that it is always interesting to see whether quantum mechanics can reveal itself on a macroscopic scale. Another reason is that the tunneling of $M(r)$ introduces a new concept, *magnetic instantons*, into the traditional field of magnetism. It turns out that the very same equation (Eq. (1.1)) that describes the classical dynamics of the magnetization also describes quantum tunneling out of metastable classical states [2]. If one switches in Eq. (1.1) from real to imaginary time, this equation possesses exact solutions, instantons [3], which carry out classically forbidden transitions of M between different equilibrium orientations. We shall see that they describe different

tunneling phenomena: quantum tunneling of M in single-domain magnetic particles [4], quantum nucleation of magnetic bubbles [5], tunneling of the Néel vector
in antiferromagnets [6], etc. Mathematically, magnetic instantons are similar to
real-time magnetic solitons, such as, e.g., domain walls, which have been studied
intensively for a long time. In contrast, intensive theoretical and experimental
research on *magnetic instantons* started only in the 1990s.

The fact that quantum mechanics can be applied at two, practically independent, levels, microscopic (the Heisenberg Hamiltonian) and macroscopic (the
Landau–Lifshitz equation), is not a unique situation for magnetism, but is a
rule in all quantum physics. Let us illustrate this point by the example of an
alpha-particle. At the most fundamental level it consists of 12 quarks interacting
via gluons. At a more macroscopic level, the alpha-particle is formed by four
nucleons interacting via pions. Finally, at the most macroscopic level, it is an elementary particle characterized by a single coordinate r. The latter description is
used in the problem of the alpha-decay of a radioactive nucleus, which is the problem of tunneling out of a metastable state. It is the strong interaction between
the quarks inside the nucleons and the strong interaction of the nucleons inside
the alpha-particle which allow the low-energy description in terms of a single
variable r. Thus, matter can be quantized at any level, provided that we do not
exceed energies at which couplings responsible for the formation of the macroscopic variable are broken and its microscopic structure becomes apparent. In
the case of the magnetic moment of a ferromagnet the relevant coupling is the
strong exchange interaction among individual spins which typically is of the
order of the Curie temperature. Consequently, for a material having the Curie
temperature of a few hundred kelvins, the low-frequency description of both
classical and macroscopic quantum dynamics, in terms of the magnetization
$M(r)$, must be rather good at and below a few kelvins.

As everyone knows from elementary quantum mechanics, heavy objects do
not tunnel; first of all, because of the very low probability of such an event. The
probability decreases as $\exp(-I_{\text{eff}}/\hbar)$ with the action, I_{eff}, associated with the
tunneling. For objects of large mass, $I_{\text{eff}} \gg \hbar$. That is why tunneling very rarely
occurs at the macroscopic level. For the magnetic moment, the quantity analogous to the mass is the effective moment of inertia associated with the rotation
of M. We shall see that in some cases it can be quite small, making the magnetization a good 'light' object for the study of macroscopic quantum tunneling
(MQT).

The term MQT was introduced by Caldeira and Leggett in their seminal work
on tunneling with dissipation [7]. Consider, e.g., tunneling of a particle of mass
m through a potential barrier $U(x)$ from a metastable energy minimum at $x = a$
to the point $x = b$ on the other side of the barrier $(U(a) = U(b) = 0)$, Fig.
1.1(a). The WKB method gives for the tunneling rate $\Gamma = A \exp(-B)$, where A
is of the order of the frequency of small oscillations, ω, near $x = a$ in Fig. 1.1(a)

and $B = 2|\int_a^b p\,\mathrm{d}x|/\hbar$. Here $p(x)$ is the imaginary linear momentum under the barrier, which satisfies $U(x) + p^2/(2m) = 0$. It is easy to see that B is the action of the mechanical motion from $x = a$ to $x = b$ in the inverted potential (Fig. 1.1(b)) or (which is the same) in imaginary time, $\tau = it(p^2 \to -p^2)$. In order of magnitude $B \simeq U_0/\omega_0$, where U_0 is the height of the barrier and ω_0 is the frequency of small oscillations at the bottom of the inverted potential in Fig. 1.1(b). In the presence of Ohmic friction the classical equation of motion is $m(\mathrm{d}^2x/\mathrm{d}t^2 + \eta\,\mathrm{d}x/\mathrm{d}t) = -\mathrm{d}U/\mathrm{d}x$. A formal switch to imaginary time gives $m(-\mathrm{d}^2x/\mathrm{d}\tau^2 + i\eta\,\mathrm{d}x/\mathrm{d}\tau) = -\mathrm{d}U/\mathrm{d}x$. The latter equation has a Fourier transform $m_{\mathrm{eff}}\omega^2 x_\omega = -(\mathrm{d}U/\mathrm{d}x)_\omega$, where $m_{\mathrm{eff}}(\omega) = m(1 + \eta/\omega)$. On the basis of this observation one can speculate that friction renormalizes the mass of the particle in the tunneling problem, making it heavier. Qualitatively, it should result in $B \to B(1 + \eta/\omega_0)$ for the tunneling exponent. This absolutely non-obvious statement has a rigorous quantum-mechanical proof [7]. It gives a powerful tool for a rough assessment of the tunneling probability based upon classical characteristics of the potential barrier and dissipative environment,

$$\Gamma \simeq \frac{\omega}{2\pi}\exp\left[-\frac{U_0}{\hbar\omega_0}\left(1 + \frac{\eta}{\omega_0}\right)\right]. \tag{1.3}$$

When one turns to the problem of magnetic MQT, a few observations are in order. First, Eq. (1.1) is very different from the conventional equation of motion for a mechanical particle discussed above. We shall see, however, that the problem of spin tunneling can be formulated in a similar way [2,4,8,9] in terms of

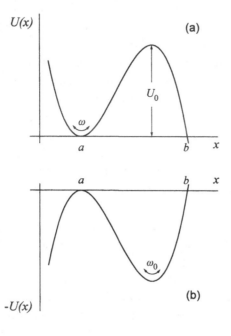

Figure 1.1 (a) Tunneling out of a metastable state. (b) Tunneling as motion in imaginary time.

the spherical coordinates, ϕ and θ, of a fixed-length spin-vector S. The corresponding energy barrier is usually due to the magnetic anisotropy. The prefactor f is the small frequency of oscillations of M in a metastable state, typically of the order of the ferromagnetic resonance (FMR) frequency. The instanton solution of Eq. (1.1) for a given tunneling problem determines ω_0 and the tunneling exponent. In most cases it can be obtained exactly. The latter is important because the rate depends exponentially on the parameters of the problem. The exact value of the prefactor is usually much more difficult to obtain, but a good estimate can be made on the basis of the arguments provided above.

Secondly, the dissipation of the motion of M is non-Ohmic and non-linear. In problems of classical magnetic dynamics it is taken into account by adding

$$\frac{\eta}{M^2} M \times \left(M \times \frac{\delta E}{\delta M} \right) \tag{1.4}$$

to the right-hand side of Eq. (1.1). One can hope that the estimate of the effect of dissipation provided by Eq. (1.3) works for the magnetic tunneling as well, but no rigorous proof of this statement exists. A general proof would not be possible because of the variety of mechanisms of dissipation in magnets. They include interactions of the magnetization with phonons, itinerant electrons, defects and nuclear spins. Some of these effects are negligible whereas others deserve attention [10, 11]. In the first approximation, the classical real-time motion of M is non-dissipative. The parameter η in Eq. (1.3) is the measure of the dissipation. It determines, e.g., the width of the ferromagnetic resonance, which can be very narrow, especially in magnetic insulators. Weak dissipation of the classical motion does not insure, however, that the dissipation has a small effect in MQT problems. Indeed, for a macroscopic variable to tunnel, the barrier must be small. In problems of magnetic tunneling this can be achieved by applying a well-controlled magnetic field that lowers the barrier. In the limit of a small barrier, as one can see from Fig. 1.1, the frequency of small oscillations near the bottom of the inverted potential, ω_0, also becomes small, and according to Eq. (1.3), the effect of dissipation may become significant.

Thirdly, any experiment is done at finite temperature, so that thermal overbarrier transitions compete with quantum tunneling. They go with the rate $\Gamma \propto \exp(-U_0/T)$. On comparing it with Eq. (1.3), one finds that, for weak and moderate dissipation, quantum tunneling dominates below $T_c \simeq \hbar\omega_0$. Thus, the characteristics of the instanton also determine the temperature range for the experimental studies of the MQT. In ferromagnets T_c can be roughly estimated as $\mu_B H_{an}$, where H_{an} is the anisotropy field owing to the term in the Hamiltonian that violates its commutation with M [4]. In antiferromagnets T_c is of the order of $\mu_B(H_{an}H_{ex})^{1/2}$, where H_{ex} is the exchange field [6]. In many materials T_c is as high as a few kelvins, which puts magnetic MQT within the temperature range easily accessible in experiments.

Magnetic materials are characterized by hysteresis phenomena arising from the existence of various metastable magnetic configurations. The simplest qualitative evidence of magnetic tunneling would be the fact that the time relaxation of the magnetic moment in a macroscopic solid continues, independently of temperature, as T goes to absolute zero. In recent years this behavior has been observed in a large number of systems [12], suggesting that magnetic tunneling is a common phenomenon that determines the low-temperature dynamics of magnetic materials. Direct comparison between theory and experiments has been hampered, however, by the fact that most experimental systems have a variety of metastable states with a wide distribution of tunneling rates. This raises the possibility that the observed temperature-independent relaxation can be explained in terms of classical dynamics taking into account the interaction between tunneling clusters (particles or domain walls), or in terms of a certain distribution of energy barriers, or in terms of self-heating arising from the release of energy from the thermal decay of metastable states. The main argument in favor of the tunneling interpretation is that any real-time dynamics of the magnetic moment, governed, e.g., by Eq. (1.1), develops on a timescale of less than 1 μs, whereas in most cases it takes hours or days for low-temperature magnetic relaxation to result in a significant change in the magnetization. Such a slow relaxation cannot be explained in terms of the real-time dynamics of M or the dynamics of the heat flow in the system. As we shall see, the most plausible explanation is provided by the exponentially slow decay of metastable states in disconnected regions of a solid.

In recent years a few experiments have been performed whose goal was to detect magnetic tunneling in a system having a very narrow distribution of barriers or no distribution at all. They included studies of an array of presumably identical single-domain particles [13], measurements of individual particles [14], and measurements of the dynamics of individual domain walls in mesoscopic wires [15]. Although these experiments have been inconclusive so far, these are very promising directions, especially in view of the rapidly advancing technology of fabricating nanometer-size particles.

Whereas the physics research on magnetic tunneling has concentrated on manufacturing mesoscopic magnets that are small enough to exhibit tunneling, chemistry has helped the problem from the other end. Magnetic molecules have been synthesized with spins as large as 10. The corresponding magnetic moment is 20 times greater than the moment of an electron. These are well-characterized objects that make possible comparison between theory and experiment free of any fitting parameters. Being intermediate between micro and macro, they allow one to study, in a macroscopic experiment, the border between MQT and the conventional quantum mechanics of a spin. The discovery of resonant spin tunneling in Mn-12 acetate [16] was an important landmark in the search for MQT of the magnetic moment.

Besides being of fundamental interest, quantum tunneling of magnetization may be important in applications. The trend of modern computer technology is toward devising smaller and smaller memory elements. The smaller the element the more sensitive it is to thermal fluctuation. The superdense databases of the future are likely to be operated at low temperature. This would suppress thermal fluctuations but not quantum transitions. The study of magnetic tunneling is, therefore, important for understanding the ultimate limitations of the miniaturization of magnetic memory elements. On the other hand, magnetic tunneling can be used for devising elements of quantum computers [17].

Chapter 2

Tunneling on a macroscopic scale

2.1   A general expression for the tunneling rate

Consider a *macroscopic* particle in the potential shown in Fig. 2.1. We shall assume that the system is prepared in a *metastable state*; that is, the initial state of the particle consists of low-lying *energy levels* near $x = 0$. We shall also think that the particle is in *equilibrium with a thermal bath* at a temperature T.

Although this picture seems physically appealing, everything in it needs explanation. First, what does it mean to state that the particle is macroscopic? We shall use this term throughout the book to express the fact that the system under consideration is large enough to behave classically during most of the time it is being observed. In other words, quantum transitions, which are forbidden by classical mechanics, are very rare. For the particle shown in Fig. 2.1 this means that the probability of its tunneling through the barrier is very small. Also the quantum-mechanical uncertainty, $(\langle x^2 \rangle)^{1/2}$, of its position around $x = 0$ (or around any other location) must be small compared with the characteristic length of the potential, say x_0. The last condition is satisfied if the ground state energy is small in comparison with the height of the barrier, U_0. This translates into the condition

$$\hbar\omega \ll U_0, \tag{2.1}$$

where ω is the frequency of small oscillations near $x = 0$. We shall also assume that the particle is macroscopic enough not to be strongly disturbed by the interaction with the measuring device; that is, the observation does not necessarily drive the particle far away from its measured position.

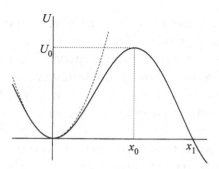

Figure 2.1 A particle in a metastable potential.

Secondly, the very existence of the metastable state means that not only the probability of quantum (underbarrier) escape of the particle into the region $x > x_1$ but also the probability of the thermal (overbarrier) transition is small. To define the transition rate Γ we need to think about the statistical ensemble of N identical metastable states that decay with time according to

$$N(t) = N_0 \exp(-\Gamma t). \tag{2.2}$$

The probability, Γ, that within a period of one second a thermal fluctuation drives the particle over the barrier is proportional to $\exp(-U_0/T)$. Thus, T must be small compared with U_0 to insure a long lifetime of the metastable state. Under this condition, the classical formula for the rate of the overbarrier transition is [18]

$$\Gamma = \frac{\omega}{2\pi} \exp\left(-\frac{U_0}{T}\right). \tag{2.3}$$

The frequency of small oscillations near the bottom of the potential well appears in the pre-exponential factor. It is called the 'attempt' frequency. A 'small probability' of escape means that the system executes many oscillations near $x = 0$ before it emerges on the other side of the barrier.

Thirdly, what is the meaning of energy levels of a metastable state? The Schrödinger equation with the potential $U(x)$ does not have the eigenstates which are localized on the left-hand side of the barrier. Apparently, we are thinking about the approximation in which the potential well to the left of the barrier is almost equivalent to the well in the case of an infinite barrier (the dashed line in Fig. 2.1). If this approximation were exact, the time evolution of eigenstates ψ_n, corresponding to energy levels E_n, would be

$$\psi_n \propto \exp(-iE_n t/\hbar). \tag{2.4}$$

Owing to the tunneling, however, these states are slowly decaying with time,

$$|\psi_n|^2 \propto \exp(-\Gamma_n t); \tag{2.5}$$

Γ_n being the rate of the escape from the level n. The simplest way to account for this decay is to add a small negative imaginary part to each level,

$$\mathrm{Im}\,(E_n) = -\frac{1}{2}\hbar\Gamma_n. \tag{2.6}$$

Finally, the condition of thermal equilibrium with the bath means that $-dU/dx$ is not the only force acting on the particle. The particle must also interact with the environment. This interaction should be large enough to bring the particle to thermal equilibrium during the lifetime of the metastable state. In terms of the thermal relaxation time, t_r, this requires that $\Gamma t_r \ll 1$. The finite value of t_r results in the $\hbar/t_r$ uncertainty in the energy of the particle. If it is greater than the spacing between the levels, $\hbar\omega$, the description of the system solely in terms of the potential $U(x)$ becomes meaningless. In this case one should explicitly take into account the environmental degrees of freedom. If, however, $\omega t_r \gg 1$ the effect of the environment reduces to the thermal distribution over well-defined levels, E_n, of finite width. This will be the case of our main interest. The two limits, $\omega t_r \gg 1$ and $\omega t_r \ll 1$, correspond in classical mechanics to weak and strong damping of the particle's oscillations near $x = 0$. The important achievement of the quantum theory (Caldeira and Leggett [7]) is that in many cases of practical interest the effect of the environment on the tunneling rate can be expressed solely in terms of the damping constant, $\eta = t_r^{-1}$. Note that, for a macroscopic system, the condition $\Gamma t_r \ll 1$ is usually satisfied independently of the relation between ω and η because of the exponential smallness of Γ. When it is not satisfied, we will talk about the underdamped regime.

Clearly, the system described above is in a mixed state that should be described in terms of the density matrix rather than the wave function. The rate of the transition out of the metastable state can be obtained by taking the Boltzmann average of Γ_n,

$$\Gamma = \frac{\sum_n \Gamma_n e^{-E_n/T}}{\sum_n e^{-E_n/T}} = \frac{2}{\hbar} \frac{\sum_n \mathrm{Im}\,(E_n)\, e^{-E_n/T}}{\sum_n e^{-E_n/T}}. \tag{2.7}$$

Here we have used the relation (2.6). The expression in the denominator of this formula is the partition function, which is related to the free energy of the system, F, through

$$Z = \sum_n e^{-E_n/T} = e^{-F/T}. \tag{2.8}$$

Writing F as

$$F = -T \ln \left[\sum_n \exp \left(-\frac{\mathrm{Re}\,(E_n) + i\,\mathrm{Im}\,(E_n)}{T} \right) \right], \tag{2.9}$$

it is easy to see that, to first order in exponentially small $\mathrm{Im}\,(E_i)$, the imaginary part of the free energy is

$$\mathrm{Im}\,(F) = \frac{\sum_n \mathrm{Im}\,(E_n)\, e^{-E_n/T}}{\sum_n e^{-E_n/T}}. \tag{2.10}$$

Comparison with Eq. (2.7) then gives the following simple formula for the escape rate [19]:

$$\Gamma = -\frac{2}{\hbar} \operatorname{Im}(F). \tag{2.11}$$

One who is familiar with the foundations of conventional statistical mechanics may feel a bit uncomfortable with the imaginary part of the free energy and even with the use of the concept of the free energy in the problem of a mechanical particle that does not possess any internal degrees of freedom. The last objection is easy to overcome by simply saying that we are considering an ensemble of identical particles distributed over the metastable energy levels. Insofar as the imaginary part of F is concerned, it should be understood formally, in terms of Eq. (2.8), via imaginary parts of E_n. One can say that this constitutes the analytical continuation of the free energy, which allows us to study the long-living metastable states along with the true thermodynamic equilibrium.

The imaginary part of Z should be small compared with the real part. Equations (2.8), (2.10), and (2.11) then give

$$\Gamma = \frac{2T}{\hbar} \frac{\operatorname{Im}(Z)}{\operatorname{Re}(Z)}. \tag{2.12}$$

The way formulas (2.11) and (2.12) have been derived suggests that they apply to a much more general class of problems than that formulated above. In fact, these formulas give the escape rate for any metastable state in quantum mechanics and quantum field theory, provided that it is small and that the system is in thermal equilibrium with the environment. We should now find the way to compute the ratio of the imaginary and real parts of the partition function in Eq. (2.12).

2.2   The path integral

In quantum theory

$$Z = \operatorname{Tr}(\rho), \tag{2.13}$$

where

$$\rho = \exp(-\mathcal{H}/T) \tag{2.14}$$

is the density matrix and $\mathcal{H}$ is the Hamiltonian of the system. In the coordinate representation

$$\rho(x_{\mathrm{f}}, x_{\mathrm{i}}) = \sum_n \phi_n(x_{\mathrm{f}}) \phi_n^*(x_{\mathrm{i}}) e^{-E_n/T}, \tag{2.15}$$

where ϕ_n are the eigenfunctions corresponding to the eigenvalues E_n. Then

$$Z = \int \mathrm{d}x \, \rho(x, x). \tag{2.16}$$

Thus, the problem of calculating Z has been reduced to the problem of calculating the sum in Eq. (2.15).

One of the possible solutions of the latter problem is based upon the formal mathematical analogy between Eq. (2.15) and the expression for the transition amplitude in quantum mechanics,

$$\langle x_f| \exp\left(-\frac{i}{\hbar}\mathcal{H}(t_f - t_i)\right)|x_i\rangle = \sum_n \phi_n(x_f)\phi_n^*(x_i) \exp\left(-\frac{i}{\hbar}E_n(t_f - t_i)\right).$$

(2.17)

It is well known [20] that this can be computed as a path integral,

$$\langle x_f| \exp\left(-\frac{i}{\hbar}\mathcal{H}(t_f - t_i)\right)|x_i\rangle$$
$$= \int_{x(t_i)=x_i}^{x(t_f)=x_f} \mathcal{D}\{x(t)\} \exp\left(\frac{i}{\hbar}\int_{t_i}^{t_f} dt\,\mathcal{L}(x, \dot{x})\right),$$

(2.18)

over all trajectories $x(t)$ that start at x_i at the moment of time t_i and end at x_f at the moment of time t_f. Here $\mathcal{L}$ is the Lagrangian of the system. For a particle of unit mass, shown in Fig. 2.1,

$$\mathcal{L} = \mathcal{L}_0 = \tfrac{1}{2}\dot{x}^2 - U(x).$$

(2.19)

For future applications we should note that a more general form of the relation (2.18) contains the path integral over both the coordinate and the momentum of the system,

$$\int \mathcal{D}\{x(t)\}\,\mathcal{D}\{p(t)\} \exp\left(\frac{i}{\hbar}\int_{t_i}^{t_f} dt\,\mathcal{L}(x, p)\right).$$

(2.20)

In the case of a mechanical particle,

$$\mathcal{L}(x, p) = p\dot{x} - \mathcal{H}(x, p),$$

(2.21)

$$\mathcal{H}(x, p) = \tfrac{1}{2}p^2 + U(x).$$

(2.22)

Then the Gaussian integration over p brings the double integral (2.20) to the single integral of Eq. (2.18). In general, however, as is also the case for the problem of tunneling of the magnetic moment, the situation can be much more complicated. For that reason, it is important to mention that the correspondence between the path integral method of calculating the propagator and the Schrödinger equation has been established only for Hamiltonians that can be presented as

$$\mathcal{H}(x, p) = \mathcal{H}_1(p) + \mathcal{H}_2(x).$$

(2.23)

We shall return to this question later when studying the problem of spin tunneling.

Let us now notice that, if one switches to the imaginary time in Eq. (2.17), by making the formal substitution $\tau = it$, then the expressions (2.15) and (2.17) become mathematically identical. It is convenient, therefore, to consider the imaginary-time version of Eq. (2.18),

$$\langle x_f| \exp\left(-\frac{1}{\hbar}\mathcal{H}(\tau_f - \tau_i)\right)|x_i\rangle$$
$$= \int_{x(\tau_i)=x_i}^{x(\tau_f)=x_f} \mathcal{D}x(\tau) \exp\left[-\frac{1}{\hbar}\int_{\tau_i}^{\tau_f} d\tau \left(\tfrac{1}{2}\dot{x}_\tau^2 + U(x)\right)\right], \tag{2.24}$$

where $\dot{x}_\tau$ is the derivative with respect to τ. This expression should be understood as the analytical continuation of Eq. (2.18) into the complex plane. The mathematical analogy then leads to the following expression for the density matrix [20]:

$$\rho = \int_{x_i}^{x_f} \mathcal{D}x(\tau) \exp\left[-\frac{1}{\hbar}\int_0^{\hbar/T} d\tau \left(\tfrac{1}{2}\dot{x}_\tau^2 + U(x)\right)\right]. \tag{2.25}$$

Correspondingly, the partition function, Eq. (2.16), becomes the integral

$$Z = \oint \mathcal{D}x(\tau) \exp\left[-\frac{1}{\hbar}\int_0^{\hbar/T} d\tau \left(\tfrac{1}{2}\dot{x}_\tau^2 + U(x)\right)\right], \tag{2.26}$$

over all $x(\tau)$ trajectories that are periodic in imaginary time with the period

$$\tau_p = \hbar/T; \tag{2.27}$$

that is, $x(0) = x(\hbar/T)$.

Although the above arguments have been known for many years, we included them for readers who are interested in fundamental aspects of the problem. There is an opinion [21] (that we share) that, besides the mathematical analogy, there should be fundamental physics, related to the properties of space–time, that explains why the transition from real time to imaginary time brings us from quantum mechanics to statistical mechanics. Theoretical and experimental studies of the problem of the decay of a metastable state may help to elucidate this physics.

2.3   Instantons

For a macroscopic particle, the imaginary time action in the exponent of Eq. (2.26) is large in comparison with $\hbar$. For that reason, only trajectories that are close to the one that minimizes the action contribute to the path integral. We shall start with $T \to 0$, which corresponds to pure quantum mechanics. In that limit the period (2.27) becomes infinite. We are looking, therefore, for closed trajectories that start at $x = 0$ at $\tau = -\infty$ and end at $x = 0$ at $\tau = +\infty$. Extremal trajectories that minimize the action satisfy

$$\ddot{x}_\tau = \frac{\partial U}{\partial x}. \tag{2.28}$$

This looks like Newton's equation of motion of the particle in the potential $-U(x)$ shown in Fig. 2.2. On the basis of this analogy, we see immediately that there are three solutions of Eq. (2.28) that satisfy the condition of periodicity:

$$x = 0,$$
$$x = x_0,$$
$$x = x_b(\tau). \tag{2.29}$$

The trivial first two solutions are just stationary points of the potential. The non-trivial solution, $x_b(\tau)$, is the so called 'bounce trajectory' or instanton [3,22]. It describes the imaginary-time motion for which the particle located at $x = 0$ starts at $\tau = -\infty$ to roll down the slope of the potential in Fig. 2.2, arrives at $x = x_1$ at $\tau = 0$, and then bounces back to $x = 0$, where it arrives at $\tau = +\infty$. The dependence of x_b on τ is shown in Fig. 2.3. It satisfies the zero-energy first integral of Eq. (2.28),

$$\tfrac{1}{2}\dot{x}_b^2 = U(x). \tag{2.30}$$

This trajectory corresponds to the tunneling of the particle from $x = 0$ to $x = x_1$. The name instanton reflects the fact that this process takes no real time, the 'motion' occurs entirely in imaginary time. Substitution of the bounce trajectory into the exponent of Eq. (2.26) gives

$$B \equiv \frac{1}{\hbar}\int d\tau \left(\tfrac{1}{2}\dot{x}_b^2 + U(x)\right) = \frac{1}{\hbar}\int d\tau\, \dot{x}_b^2 = \frac{2}{\hbar}\int_0^{x_1} dx\, (2U(x))^{1/2}. \tag{2.31}$$

The last expression is the familiar WKB exponent for one-dimensional tunneling.

Trajectories that are close to the bounce trajectory give comparable contributions to the path integral. Consider perturbations of the bounce,

$$x(\tau) = x_b(\tau) + \delta x(\tau), \tag{2.32}$$

where $\delta x(\tau) = \delta x(\tau + \tau_p)$. Substitution into the exponent of Eq. (2.26) gives to second order in $\delta x(\tau)$

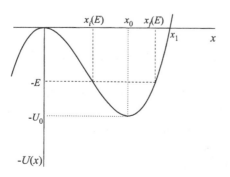

Figure 2.2 The inverted potential, $-U(x)$.

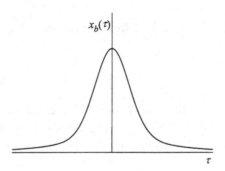

Figure 2.3 The τ-dependence of the instanton.

$$\frac{1}{\hbar}\oint d\tau \left(\tfrac{1}{2}\dot{x}_b^2 + U(x)\right) = B + \frac{1}{2\hbar}\int d\tau\,\delta x[-\partial_\tau^2 + \partial_x^2 U(x = x_b)]. \qquad (2.33)$$

Let us now expand δx into a series of eigenfunctions of the operator $-\partial_\tau^2 + \partial_x^2 U(x = x_b)$,

$$\delta x(\tau) = \sum_n C_n x_n(\tau), \qquad (2.34)$$

where

$$\int d\tau\, x_n(\tau)x_m(\tau) = \delta_{nm}. \qquad (2.35)$$

This gives

$$\frac{1}{\hbar}\int d\tau \left(\tfrac{1}{2}\dot{x}_b^2 + U(x)\right) = B + \frac{1}{2\hbar}\sum_n \lambda_n C_n^2, \qquad (2.36)$$

where λ_n is an eigenvalue corresponding to x_n.

To compute the integral (2.26) over these trajectories we must first perform integration over C_n,

$$\begin{aligned}
\mathrm{e}^{-B}\prod_n \int_{-\infty}^{+\infty} &\frac{dC_n}{(2\pi\hbar)^{1/2}} \exp\left(-\frac{\lambda_n C_n^2}{2\hbar}\right) \\
&= \mathrm{e}^{-B}\prod_n \lambda_n^{-1/2} = \mathrm{e}^{-B}\{\det[-\partial_\tau^2 + \partial_x^2 U(x = x_b)]\}^{-1/2}.
\end{aligned} \qquad (2.37)$$

The chosen measure of the integration, $2\pi\hbar$, need not be questioned because it will cancel out from the ratio of the imaginary and real parts of the partition function in Eq. (2.12). A tricky point, however, arises if one notices that we can arbitrarily displace the bounce trajectory in imaginary time; that is, $x_b(\tau + \tau_0)$, with $0 < \tau_0 < \hbar/T$, is another extremal trajectory. The perturbation of the bounce that corresponds to this translational invariance in imaginary time does not cost any energy, so that the corresponding eigenvalue, say λ_1, is zero. It is called the zero mode of the instanton. Because of it, the expression (2.37) is formally infinite, for it has $\lambda_1^{1/2}$ in the denominator. We should correct this situation by noticing that the integral over C_1, which leads to the infinity, must be equivalent to the integration over τ_0,

$$\int_0^{\hbar/T} d\tau_0 = \hbar/T. \qquad (2.38)$$

Thus, the origin of the infinity is in the limit of $T \to 0$. This is encouraging because the transition rate of Eq. (2.12) has T in the numerator, which must cancel the divergency. We now only need to properly switch from the integration over C_1 to the integration over τ_0.

The zero mode satisfies

$$\ddot{x}_1 = V''(x_b)x_1. \qquad (2.39)$$

With the help of Eq. (2.30) it is easy to see that the solution is

$$x_1 = k\dot{x}_b. \qquad (2.40)$$

For the constant k the normalization condition gives

$$\int d\tau\, x_1^2 = k^2 \int d\tau\, \dot{x}_b^2 = k^2 \hbar B = 1. \qquad (2.41)$$

By noticing that, for the zero mode,

$$dx = \dot{x}_b\, d\tau_0 = x_1\, dC_1, \qquad (2.42)$$

we obtain

$$dC_1 = (\dot{x}_b/x_1)\, d\tau_0 = (\hbar B)^{1/2}\, d\tau_0. \qquad (2.43)$$

Consequently, we should exclude the $\lambda_1 = 0$ eigenvalue from the expression (2.37) and replace it by the factor

$$\int \frac{dC_1}{(2\pi\hbar)^{1/2}} = \left(\frac{B}{2\pi}\right)^{1/2} \int d\tau_0 = \left(\frac{B}{2\pi}\right)^{1/2} \frac{\hbar}{T}. \qquad (2.44)$$

Our troubles have not ended, however. Indeed, looking at the shape of the bounce, Fig. 2.3, and Eq. (2.40), one immediately notices that the zero mode has the shape shown in Fig. 2.4; that is, it has a node. Thus, as we know from conventional quantum mechanics, $\lambda_1 = 0$ cannot be the minimal eigenvalue of

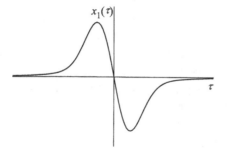

$x_1(\tau)$

τ

Figure 2.4 The τ-dependence of the zero mode.

the operator $-\partial_\tau^2 + \partial_x^2 U(x = x_b)$. The minimal eigenvalue, $\lambda_0 < 0$, must correspond to a nodeless eigenfunction, $x_0(\tau)$. Formally, it makes the determinant of Eq. (2.37) imaginary, which we need, according to Eq. (2.12), in order to produce the imaginary part of the partition function. From the mathematical point of view, however, for a negative λ_0, the integral (2.37) over C_0 diverges. The only way out of this trouble is to shift the integration over C_0 to the positive imaginary axis in the complex plane. Finally we obtain for the imaginary part of the partition function

$$\mathrm{Im}\,(Z) = \frac{\hbar}{2T} \left(\frac{B}{2\pi}\right)^{1/2} e^{-B} |\{\det' [-\partial_\tau^2 + \partial_x^2 U(x = x_b)]\}|^{-1/2}, \qquad (2.45)$$

where $\det'$ means the omission of $\lambda_1 = 0$.

To obtain the tunneling rate we should now compute the real part of Z. This is easy, if one recalls that we are interested in the case in which tunneling is a very rare event, that is, $\mathrm{Im}\,(Z) \ll \mathrm{Re}\,(Z)$. The real part of Z must be, therefore, dominated by trajectories near $x = 0$ that correspond to the long-living metastable state. For this state $B = 0$, and no zero mode exists. Then, the identical calculation without the problems encountered for $\mathrm{Im}\,(Z)$ gives for the real part

$$\mathrm{Re}\,(Z) = \{\det [-\partial_\tau^2 + \partial_x^2 U(x = 0)]\}^{-1/2}. \qquad (2.46)$$

We have not yet discussed another stationary point, $x = x_0$. It is easy to see, however, that, at $T = 0$, it does not contribute to the path integral because the corresponding value of B is infinite. This will change at $T \neq 0$, as we shall see in the next section.

Substitution of Eqs. (2.45) and (2.46) into Eq. (2.12) finally gives [22]

$$\Gamma = \left(\frac{B}{2\pi}\right)^{1/2} e^{-B} \left| \frac{\det [-\partial_\tau^2 + \partial_x^2 U(x = 0)]}{\det' [-\partial_\tau^2 + \partial_x^2 U(x = x_b)]} \right|^{1/2}. \qquad (2.47)$$

Because one eigenvalue, of dimensionality s^{-2}, has been omitted from the lower determinant, it is easy to see that Γ has the correct dimensionality of s^{-1}, that is, the number of transitions per second. $B \gg 1$ is needed to insure the smallness of Γ, which is the condition of validity of our quasiclassical approximation.

2.4 Finite-temperature effects

Let us now consider temperature effects. According to the general rule derived in Section 2.2 we must compute the path integral (2.26) over trajectories that are periodic in imaginary time with the period $\hbar/T$. Just like previously, the integral must be dominated by the vicinities of extremal trajectories given by Eq. (2.28). Stationary solutions $x = 0$ and $x = x_0$ satisfy the periodicity condition automatically. The bounce trajectory, however, must be selected such that it has the required period. It is easy to see, just by looking at Fig. 2.2, that there are periodic trajectories that correspond to the imaginary-time oscillations in the inverted potential between $x_i(E)$ and $x_f(E)$. These trajectories satisfy the first integral of Eq. (2.28) at a certain energy, $E < U_0$,

$$\tfrac{1}{2}\dot{x}_\tau^2 = U(x) - E. \tag{2.48}$$

The energy must be selected under the condition that it produces the correct period,

$$\tau_p(E) = \frac{\hbar}{T} = \sqrt{2} \int_{x_i(E)}^{x_f(E)} \frac{dx}{(U(x) - E)^{1/2}}. \tag{2.49}$$

To distinguish these trajectories from instantons (which correspond to $T = 0, E = 0, \tau_p = \infty$) we shall call them 'thermons' [23].

The transition rate can be written as

$$\Gamma(T) = A(T)e^{-B(T)}. \tag{2.50}$$

Here the exponent is given by

$$B(T) = \frac{1}{\hbar} \int_0^{\hbar/T} d\tau \left(\tfrac{1}{2}\dot{x}_\tau^2 + U(x)\right), \tag{2.51}$$

$x(\tau)$ being the extremal periodic trajectory at a given temperature T. The pre-exponential factor, $A(T)$, can be obtained by making a small perturbation around that trajectory, in the same manner as we did at $T = 0$. Note that Eq. (2.28) may have a few solutions with the required period, each of them contributing to the path integral. Because of the exponential dependence of the rate on B, however, a high accuracy can be obtained by selecting the extremal trajectory that corresponds to the absolute minimum of B. This should be true for all T, except in the vicinity of a certain temperature that marks the crossover from thermally assisted tunneling to purely thermal activation (see below).

For a wide class of potentials, the period of a thermon is a monotonic function of the energy. Potentials of the form $U(x) = -x^2 + x^3$ and $U(x) = -x^2 + x^4$ belong to that class. A typical dependence of τ_p on E is shown in Fig. 2.5. The period becomes infinite in the limit of zero energy, which corresponds to the instanton. In the opposite limit, $E \to U_0$, it tends to a finite value, $\tau_0 = 2\pi/\omega_0$, where ω_0 is the frequency of small oscillations near the bottom of the inverted potential, Fig. 2.2. In that limit the thermon reduces to

$$x(\tau) = x_0 + a(E) \sin(\omega_0\tau). \tag{2.52}$$

Consequently, there exists a maximal temperature,

$$T_0 = \frac{\hbar}{\tau_0} = \frac{\hbar\omega_0}{2\pi}, \tag{2.53}$$

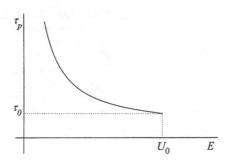

Figure 2.5 The mono-
tonic dependence of the
period on energy.

above which there is no thermon with the required period. Above T_0, the only extremal trajectories that satisfy the condition of periodicity are the stationary points $x = 0$ and $x = x_0$. The first of them, as we know, just normalizes the transition rate; the second, according to Eq. (2.26), produces the exponent,

$$B_0 = U_0/T. \tag{2.54}$$

Thus, quite remarkably, the same expression (2.26) produces the tunneling exponent at $T = 0$ and the Boltzmann exponent at high temperature.

By using Eq. (2.49) it is easy to show that

$$\frac{\mathrm{d}B_0}{\mathrm{d}\tau_\mathrm{p}} = U_0, \quad \frac{\mathrm{d}B_T}{\mathrm{d}\tau_\mathrm{p}} = \frac{E}{\hbar} > 0, \tag{2.55}$$

where B_T denotes the thermon's action. These formulas allow one to analyze the temperature dependence of the transition exponent on the basis of the dependence of τ_p on E. On substituting (2.52) into the exponent of Eq. (2.26), one can easily see that, for $\tau_\mathrm{p}(E)$ shown in Fig. 2.5, $B_T < B_0$ at $T < T_0$. Because the amplitude of the thermon, $a(E)$, becomes zero at $E = U_0$, the thermon's action coincides with the thermal action at $T = T_0$, that is, $B_T(T_0) = B_0$. Equation (2.55) also gives

$$\left(\frac{\mathrm{d}B_T}{\mathrm{d}T}\right)_{T=T_0} = \left(\frac{\mathrm{d}B_0}{\mathrm{d}T}\right)_{T=T_0}. \tag{2.56}$$

The temperature dependence of the thermon's action, B_T, together with $B_0(T)$, are shown in Fig. 2.6. According to the above analyses, the two actions, corresponding to the thermally assisted tunneling and to purely thermal activation, respectively, join smoothly at $T = T_0$.

As we shall see presently, this is not necessarily true for all potentials, even those of a smooth shape as shown in Fig. 2.1 [23]. To illustrate this point, consider a potential that is rather flat (that is, changes slowly) near the bottom ($x = 0$) and the top ($x = x_0$) but is steep in between. For such a potential $\tau_\mathrm{p}(E)$ may have a minimum at some $E_1 < U_0$, as shown in Fig. 2.7. Figure 2.8 shows the corresponding dependence of the thermon's action on temperature. As the temperature is lowered the thermon's action becomes smaller than the thermal action at some temperature T_c satisfying

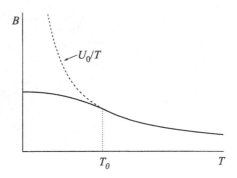

Figure 2.6 The temperature dependence of the absolute minimum of B corresponding to $\tau(E)$ of Fig. 2.5. The solid line represents the thermon's action; the dashed line corresponds to purely thermal activation.

$$T_0 < T_c < T_1, \tag{2.57}$$

where T_0 is defined by Eq. (2.53) and

$$T_1 = \frac{\hbar}{\tau_p(E_1)}. \tag{2.58}$$

It is clear that now the first derivative of the absolute minimum of B on temperature is discontinuous at T_c.

One may notice the analogy with phase transitions, $B(T)$ resembling the behavior of the thermodynamic potential and $\mathrm{d}E/\mathrm{d}T$ being analogous to the specific heat. The situation shown in Fig. 2.6 may then be called the second-order transition from tunneling to thermal activation, whereas the situation shown in Fig. 2.8 is analogous to the first-order transition. This analogy, however, only exists before we take into account quantum corrections to the transition rate and compute the determinants in the same manner as at $T = 0$. A full analysis of the 'first-order' transition is lacking at the moment. The analysis has been done [24] for the 'second-order' transition. The calculation of determinants is laborious but rewarding. It shows that both the exponent $B(T)$ and the pre-exponential factor $A(T)$ are smooth functions of T, producing the tunneling rate of Eq.

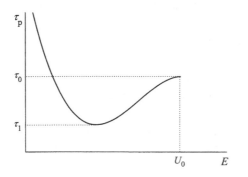

Figure 2.7 The non-monotonic dependence of the period on the energy.

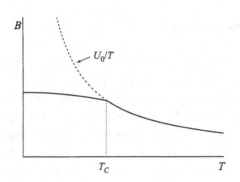

Figure 2.8 The temperature dependence of the thermon's action (solid line) corresponding to $\tau(E)$ of Fig. 2.7. The dashed line corresponds to purely thermal activation.

(2.47) at $T = 0$ and the classical expression (2.3) at high temperature. The smoothening around T_0 takes place within a temperature range $\delta T / T_0 \simeq 1/B^{1/2}$, which is rather narrow for $B \gg 1$. We shall see in Chapter 7 that both the 'first-order' and the 'second-order' transitions exist in the problem of spin tunneling.

Throughout this book our main interest will be the study of the transition exponent. The prefactor is not only much more difficult to compute but also difficult to extract from experimental data, at least when it comes to magnetic tunneling. From the experimental point of view, it is convenient to express the tunneling rate in the form

$$\Gamma = A(T) \exp\left(-\frac{U_0}{T_{\text{esc}}(T)}\right), \tag{2.59}$$

where T_{esc} is the characteristic 'escape temperature'. A typical dependence of T_{esc} on T in the presence of quantum tunneling is shown in Fig. 2.9. At high temperature $T_{\text{esc}} = T$, that is the decay of a metastable state is due to thermal overbarrier transition. When tunneling is present at low temperature T_{esc} tends to a nonzero constant.

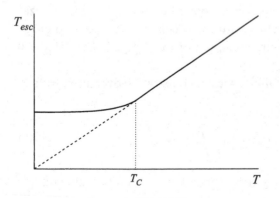

Figure 2.9 The temperature dependence of the 'escape temperature' in the presence of quantum tunneling.

2.5   Tunneling with dissipation

Up to this moment the only effect of the thermal bath on the transition rate came through the temperature, that is, through the fact that the particle can go through (or over) the barrier from excited levels that it occupies with a certain probability. There is another effect of the environment, however, which has not been taken into account yet. Interaction with microscopic degrees of freedom changes the dynamics of the particle. It results in the dissipation of the particle's motion in real time. If the friction is linear with respect to the velocity, the corresponding equation of motion becomes

$$\ddot{x}_t + \eta \dot{x}_t + U'_x = 0, \tag{2.60}$$

where η is the coefficient of friction. The important achievement of the quantum theory [7] is that in a rather general case the effect of the dissipation on the tunneling rate can be expressed entirely in terms of the coefficient η.

We shall start with a microscopic Lagrangian

$$\mathcal{L} = \mathcal{L}_0 + \mathcal{L}_{env} + \mathcal{L}_{int}. \tag{2.61}$$

Here $\mathcal{L}_0$ is given by Eq. (2.19), $\mathcal{L}_{env}$ is the Lagrangian of the environment, and $\mathcal{L}_{int}$ represents the interaction between the particle and the environment. It is close enough to the real-life situation to choose $\mathcal{L}_{env}$ as a sum over harmonic oscillators,

$$\mathcal{L}_{env} = \sum_i (\tfrac{1}{2} m_i \dot{x}_i^2 - \tfrac{1}{2} m_i \omega_i^2 x_i^2), \tag{2.62}$$

where m_i, ω_i, and x_i are the mass, the frequency, and the coordinate of the ith oscillator. For $\mathcal{L}_{int}$ we shall choose the simplest form of the interaction, linear in x and x_i, just to illustrate the point,

$$\mathcal{L}_{int} = -\sum_i C_i x_i x - \sum_i \frac{C_i^2 x^2}{2 m_i \omega_i^2}, \tag{2.63}$$

where the last term on the right-hand side is required to cancel the renormalization of the potential by the interaction with the oscillators. It is easy to check that, owing to this term, the absolute minimum of the energy is kept at $x = 0$ and $x_i = 0$.

The partition function now contains functional integration over all degrees of freedom,

$$Z = \prod_i \oint \mathcal{D}x_i(\tau) \oint \mathcal{D}x(\tau) \exp\left(-\frac{1}{\hbar} \int_0^{\hbar/T} d\tau \, \mathcal{L}_E\right), \tag{2.64}$$

where $\mathcal{L}_E$ is the Euclidean (imaginary-time) version of the total Lagrangian,

$$\mathcal{L}_E = \tfrac{1}{2}\dot{x}^2 + U(x) + \tfrac{1}{2}\sum_i m_i(\dot{x}_i^2 + \omega_i^2 x_i^2) + x\sum_i C_i x_i + x^2 \sum_i \frac{C_i^2}{2m_i\omega_i^2}.$$

$$(2.65)$$

The Gaussian integration over x_i then gives [7]

$$Z = \int \mathcal{D}x(\tau)\, e^{-I_{\mathrm{eff}}/\hbar},$$

$$(2.66)$$

where

$$I_{\mathrm{eff}} = \int_0^{\hbar/T} d\tau \left(\tfrac{1}{2}\dot{x}_\tau^2 + U(x)\right)$$

$$+ \tfrac{1}{2}\int_{-\infty}^{+\infty} d\tau' \int_0^{\hbar/T} d\tau\, \alpha(\tau - \tau')[x(\tau) - x(\tau')]^2$$

$$(2.67)$$

$$\alpha(\tau - \tau') = \sum_i \frac{C_i^2}{4m_i\omega_i}\, e^{-\omega_i|\tau - \tau'|}.$$

$$(2.68)$$

We want to express α in terms of the friction coefficient η. For that purpose let us go back to the real-time equations of motion which follow from the total Lagrangian,

$$\ddot{x} = -\frac{dU}{dx} - \sum_i \frac{C_i^2}{m_i\omega_i^2} x - \sum_i C_i x_i$$

$$m_i\ddot{x}_i = -m_i\omega_i^2 x_i - C_i x.$$

$$(2.69)$$

Taking the Fourier transform of these equations and using the second equation to express x_i via x, we obtain the following equation for x_ω:

$$-\omega^2 x_\omega - K(\omega)x_\omega + \left(\frac{dU}{dx}\right)_\omega = 0,$$

$$(2.70)$$

where

$$K(\omega) = \sum_i \frac{C_i^2\omega^2}{m_i\omega_i^2(\omega_i^2 - \omega^2)}.$$

$$(2.71)$$

To introduce dissipation we must add a small imaginary part to the frequency, $\mathrm{Im}\,(\omega) \to 0$. Then, to leading order in $\mathrm{Im}\,(\omega)$,

$$\mathrm{Im}\,[K(\omega)] = \sum_i \frac{2C_i^2\omega\,\mathrm{Im}\,(\omega)}{m_i[(\omega_i^2 - \omega^2)^2 + 4\omega^2\,\mathrm{Im}\,^2\omega]},$$

$$(2.72)$$

where by ω we understand its real part. Finally, using the mathematical relation

$$\lim_{\epsilon \to 0} \frac{\epsilon}{\beta^2 + \epsilon^2} = \pi\delta(\beta),$$

$$(2.73)$$

we obtain in the limit of Im $(\omega) \to 0$

$$\text{Im}\,[K(\omega)] = \frac{\pi}{2} \sum_i \frac{C_i^2}{m_i \omega_i} \delta(\omega - \omega_i). \tag{2.74}$$

We shall now consider the limit of small frequencies when the classical equation (2.60) with a phenomenological friction constant is applied. This equation, in the limit of small ω, must coincide with Eq. (2.70). In that limit the real part of $K(\omega)$, according to Eq. (2.71), disappears, while the imaginary part, according to Eq. (2.74), remains finite. This gives

$$K(\omega) = i\eta\omega, \tag{2.75}$$

where

$$\eta = \frac{\pi}{2} \sum_i \frac{C_i^2}{m_i \omega_i^2} \delta(\omega - \omega_i). \tag{2.76}$$

We thus have the expression for the macroscopic friction coefficient η in terms of microscopic oscillators. On comparing it with Eq. (2.68), we see that

$$\alpha = \frac{1}{2\pi} \int_0^\infty d\omega\, \eta\omega e^{-\omega|\tau - \tau'|} = \frac{\eta}{2\pi} \frac{1}{|\tau - \tau'|^2}. \tag{2.77}$$

Substitution of α into Eq. (2.67), finally, gives the Caldeira–Leggett effective action for quantum mechanics with dissipation [7],

$$I_{\text{eff}} = \int_0^{\hbar/T} d\tau \left(\tfrac{1}{2}\dot{x}_\tau^2 + U(x)\right) + \frac{\eta}{4\pi} \int_{-\infty}^{+\infty} d\tau' \int_0^{\hbar/T} d\tau\, \frac{[x(\tau) - x(\tau')]^2}{(\tau - \tau')^2}. \tag{2.78}$$

This is a simple and beautiful formula that allows one to develop a quasiclassical approximation to quantum mechanics on the basis of the classical dissipative dynamics of Eq. (2.60). Of course, we should remember all of the simplifying assumptions which went into the derivation of this formula. The most restrictive of them is the linear (in environmental coordinates x_i) coupling of the tunneling variable to the microscopic degrees of freedom.

To obtain the tunneling rate, one must now repeat the mathematics of Sections 2.3 and 2.4 for the effective action (2.78). The rate again can be written as $\Gamma = A \exp(-B)$. Here B is the value of $I_{\text{eff}}/\hbar$ determined by $x(\tau)$ that corresponds to the absolute minimum of I_{eff}. The pre-exponential factor A is given by

$$A = \left(\frac{B}{2\pi}\right)^{1/2} \left|\frac{\det \hat{D}_0}{\det' \hat{D}_1}\right|^{1/2}, \tag{2.79}$$

where

$$\hat{D}_0 x(\tau) = [-\partial_\tau^2 + U(x=0)]x(\tau) + \frac{\eta}{\pi}\int_0^{\hbar/T} d\tau' \frac{x(\tau)-x(\tau')}{(\tau-\tau')^2}$$

$$\hat{D}_1 x(\tau) = [-\partial_\tau^2 + U(x=x_b)]x(\tau) + \frac{\eta}{\pi}\int_0^{\hbar/T} d\tau' \frac{x(\tau)-x(\tau')}{(\tau-\tau')^2}. \qquad (2.80)$$

It may be instructive to write Eq. (2.78) in terms of the Fourier transform of the coordinate of the particle of mass m. At $T=0$ the answer reads [25]

$$I_{\text{eff}} = \tfrac{1}{2}\pi \int_{-\infty}^{+\infty} d\omega\, \tfrac{1}{2} m_{\text{eff}} \omega^2 |x_\omega|^2 + I_U, \qquad (2.81)$$

where S_U is the part of the effective action that depends on the potential $U(x)$ and

$$m_{\text{eff}} = m + \frac{\eta}{|\omega|}. \qquad (2.82)$$

The finite-temperature version of this equation is

$$I_{\text{eff}} = \frac{T}{\hbar}\int_0^{\hbar/T} d\tau \int_0^{\hbar/T} d\tau'\, \dot{x}(\tau)\dot{x}(\tau') \sum_{n=1}^{\infty} m_{\text{eff}}^{(n)}(T) \cos\left(\frac{2\pi n T(\tau-\tau')}{\hbar}\right) + I_U, \qquad (2.83)$$

where

$$m_{\text{eff}}^{(n)}(T) = m + \frac{\hbar\eta}{2\pi n T}. \qquad (2.84)$$

It follows from Eqs. (2.82) and (2.83) that a rough estimate of the effect of dissipation on the tunneling exponent can be given in terms of the bare exponent B_0, the characteristic instanton frequency ω_0, and the friction constant η,

$$B \simeq B_0\left(1 + \frac{\eta}{\omega_0}\right). \qquad (2.85)$$

This is a remarkable and non-trivial observation [7]. Note that the prefactor in problems with dissipation needs much more work, which is worthwhile only in cases in which precise comparison between theory and experiment is possible.

2.6   Generalization to quantum field theory

In this section we shall briefly describe the generalization of the above method to problems of quantum field theory. Consider a scalar field $\phi(r, t)$ in $3 + 1$ dimensions, described by the Lagrangian

$$\mathcal{L} = \tfrac{1}{2}(\partial_t \phi)^2 - \tfrac{1}{2}(\nabla\phi)^2 - U(\phi), \qquad (2.86)$$

where the potential $U(\phi)$ has the shape shown in Fig. 2.10. We shall assume that the initial state is $\phi = 0$ in the entire space. This is a metastable state that decays via nucleation of a finite-size bubble with $\phi \neq 0$. The bubble then expands, lead-

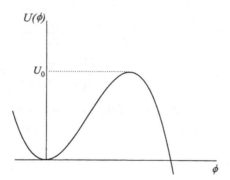

Figure 2.10 The shape of the potential $U(\phi)$.

ing to the onset of a new phase. We are only interested in the lifetime of a metastable state, not in the final equilibrium state.

Using the same method as in Section 2.2, one can prove that the general formula (2.12) for the rate holds for this case as well. We must, therefore, compute the partition function

$$Z = \oint \mathcal{D}\{\phi(\mathbf{r}, t)\} \exp\left[-\frac{1}{\hbar} \int d\tau \int d^3 r \left(\tfrac{1}{2}(\partial_\tau \phi)^2 + \tfrac{1}{2}(\nabla \phi)^2 + U(\phi)\right)\right].$$

$$(2.87)$$

Here the path integral is over all $\phi(\mathbf{r}, t)$ configurations that are periodic in imaginary time with the period $\tau_p = \hbar/T$. It is clear that, at high temperature, the rate must be dominated by thermal effects whereas at low temperature the nucleation may occur via quantum tunneling.

The extremal field configurations which minimize the action in the exponent of Eq. (2.83) satisfy

$$\ddot{\phi} + \nabla^2 \phi = U'(\phi).$$

$$(2.88)$$

Figure 2.11 The shape of the instanton at $T = 0$.

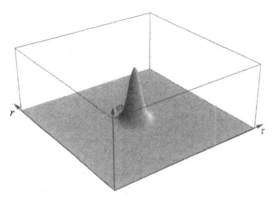

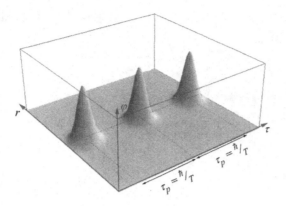

Figure 2.12 An instanton gas at $T \ll T_c$.

Consider first $T = 0$, when the period is infinite. It has been shown [22] that the action in this case is minimized by the solution of Eq. (2.84) which has a spherical symmetry, $\phi = \phi(r^2 + \tau^2)$, where $r^2 = x^2 + y^2 + z^2$. Its typical shape is shown in Fig. 2.11. For reasonable potentials, ϕ decreases exponentially far from the center. In general, analytical solutions of Eq. (2.84) in more than $1 + 1$ dimensions are impossible to obtain. The instanton must be computed numerically. Its substitution into the imaginary-time action then gives the tunneling exponent. To obtain the prefactor one must study space–time perturbations of the instanton and compute the determinants in the spirit of Section 2.3. This is usually not an easy task, even from the numerical point of view, because one should take zero and negative modes into account.

Let us now turn to the case of low temperature, $T \ll T_c$, where T_c is the temperature of the crossover from the quantum to the thermal regime. In this case, the periodic in τ solution of Eq. (2.84) can be approximated by the infinite chain of $T = 0$ instantons arranged periodically on the τ-axis as shown in Fig. 2.12. At low enough temperature τ_p is large, so that the overlap between neighboring instantons is exponentially small. This can be called the instanton gas approximation. Note that the integration with respect to τ in Eq. (2.83) is over one

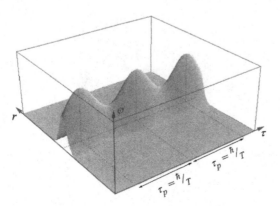

Figure 2.13 An instanton liquid at $T < T_c$.

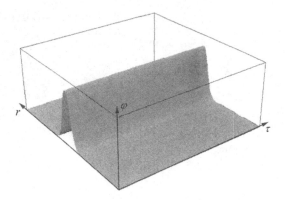

Figure 2.14 An instanton solid at $T > T_c$.

period. One should expect, therefore, that, at low temperature, thermal corrections to the tunneling rate are exponentially small.

The effect of temperature becomes noticeable when instantons overlap strongly, as illustrated in Fig. 2.13. This structure of the extremal configuration can be called the instanton liquid. For a given potential the corresponding periodic solution of Eq. (2.84) must be found numerically. The corresponding minimum of the Euclidean action gives the temperature dependence of the transition exponent under the regime of thermally assisted tunneling.

As the temperature continues to rise and the period in imaginary time continues to decrease, one finds that, at some temperature, T_c, the absolute minimum of the Euclidian action switches to the τ-independent solution shown in Fig. 2.14. This can be called the instanton solid. This solution is similar to the stationary point $x = x_0$ in the case of a mechanical particle. It satisfies $\nabla^2 \phi = U'(\phi)$ and gives the shape of the critical nucleus in the thermodynamic theory of nucleation. Its substitution into the action produces $B = U_0/T$, where U_0 is the energy barrier for the nucleation. At $T = T_c$, therefore, the transition from quantum to thermal activation manifests itself. Later we shall apply this technique to the problem of the nucleation of magnetic bubbles.

Chapter 3

Magnetism in imaginary time

3.1   An overview of the problem of spin tunneling

In this chapter we shall apply conceptions from the previous chapter to tunneling of the magnetic moment. We will begin with the problem of spin tunneling in a nanometer-size single-domain ferromagnetic particle. The particle will be considered small enough that one can treat its moment as a fixed-length vector that can only rotate as a whole. This approximation, which should be good in some but not all cases, is discussed in some detail below.

The absolute value of the magnetization in ferromagnets, M_0, defined as the magnetic moment of $1 \, cm^3$ of the material, is formed by the strong exchange interaction between individual spins. This interaction has its origin in the non-diagonal (in spin) matrix elements of the Coulomb interaction between electrons of neighboring atoms. For that reason the energy, ϵ_{ex}, required to rotate one atomic spin with respect to its neighbors is of the order of 1 eV times a factor proportional to the overlap of the corresponding wave functions. A good estimate of that energy is provided by the Curie temperature. If one takes a material that is ferromagnetic at room temperature and cools it down to a few kelvins, the probability that any individual atomic spin flips with respect to other spins is proportional to $\exp\left(-\epsilon_{ex}/T\right)$, that is, exponentially small. For that reason, at low temperature, treating the magnetization of a bulk ferromagnet as a fixed-length vector must always be a good approximation. Consider now rotation of this vector as a whole. Although not changing the exchange interaction between the spins, it does require, however, a certain energy that is called the energy of the magnetic anisotropy. This comes from the fact that, in a crystalline ferromagnet, the magnetic energy is minimized when the magnetic moment is directed along certain axes determined by the symmetry of the crystal. For instance, in cobalt, it is the hexagonal axis of the crystal, whereas in iron there are three cubic axes along which this occurs. The energy of the magnetic anisotropy, ϵ_{an}, is related to the orbital motion of electrons and has a relativistic factor, $(v/c)^2$, compared with the exchange energy. Typically, $\epsilon_{an}/\epsilon_{ex} \simeq 10^{-5}$–$10^{-3}$.

It is energetically favorable for bulk ferromagnets to split into magnetic domains in order to decrease the energy of the magnetic field surrounding the magnet. The average size of domains depends on the size and the shape of the sample. Typical domains are larger than 1 μm. They are separated by domain walls that are much narrower. Although there are materials with domain walls as narrow as the distance between the atomic planes, a, most ferromagnets' domain walls are thick compared with that distance. A typical wall is a few hundred atomic spacings thick. Inside such a wall the magnetization smoothly rotates from one domain to another. The width of the domain wall λ is determined by the balance between the exchange energy and the anisotropy energy and is given by $\lambda \simeq (\epsilon_{ex}/\epsilon_{an})^{1/2} a$. It is the high exchange energy that makes difficult any non-uniform rotation of the magnetization on a scale less than λ.

On the basis of the above picture, one should expect particles of size $d \ll \lambda$ to be uniformly magnetized. This is confirmed by experiment. To find the condition of the uniform magnetization more accurately, one should take into acount the magnetic dipole interaction between the atomic moments of the particle. This energy can be written as $\epsilon_d = 2\pi N_{ik} M_i M_k$, where M is the magnetic moment of the particle and $\hat{N}$ is a tensor that depends on the shape of the particle. Besides, for a sphere, ϵ_d depends on the orientation of M and, thus, effectively contributes to the magnetic anisotropy. If one distinguishes between the anisotropy energy that originates from the crystalline structure, ϵ_{an}, and the anisotropy energy coming from the dipole interaction, ϵ_d, the condition for a particle of diameter d to be a monodomain one can be written in the form [26] $d < d_0$, where d_0 is given by

$$d_0 = \left(\frac{\epsilon_{an}}{\epsilon_d}\right)^{1/2} \lambda \tag{3.1}$$

for a weak magnetocrystalline anisotropy ($\epsilon_{an} \ll \epsilon_d$), and by

$$d_0 = \left(\frac{\epsilon_{an}}{\epsilon_d}\right) \lambda \tag{3.2}$$

for a strong magnetocrystalline anisotropy ($\epsilon_{an} \gg \epsilon_d$). The weak-anisotropy case applies, e.g., to crystalline iron, for which $\lambda \simeq 50$ nm and $d_0 \simeq 15$ nm, whereas an example of the strong-anisotropy case would be MnBi, for which $\lambda \simeq 10$ nm and $d_0 \simeq 500$ nm. Notice that, in the strong-anisotropy case, even particles large compared with the domain wall thickness are uniformly magnetized.

In all the above considerations we assumed that the chemical and structural composition of the particle was uniform. In most cases this is not true. For instance, many magnetic particles develop an oxide layer on the surface, which can be non-magnetic. It is important to understand, therefore, that what we call the size or the volume of the particle refers to its magnetic core and may, in fact, be smaller than the real size. Even when the particle has a uniform atomic structure, spins at the surface have different interactions than spins in the bulk. It is possible, therefore, that the condition of uniform magnetization breaks at the surface. This could be the case if at the interface between the particle and the matrix, in which it is embedded, some of the ferromagnetic couplings were to switch to antiferromagnetic couplings. In such a case the frustrating interactions at the surface could result in some kind of spin-glass ordering of surface spins. At the time when this book was being written, experimental data that would confirm such a picture were lacking. A few numerical studies [27, 28] used very weak ferromagnetic exchange in order to amplify the effects of magnetic non-uniformity. In strong ferromagnets, however, the exchange interaction at the surface, unless it is significantly altered by the matrix, should be a noticeable fraction of the exchange in the bulk. One should expect, therefore, the condition of uniform magnetization of the magnetic core of the particle to apply to many strong ferromagnets.

In the absence of a magnetic field, the magnetic energy of the particle satisfies $E(M) = E(-M)$, independently of the shape of the particle. This is a consequence of the symmetry with respect to time reversal: M changes its sign under the transformation $t \rightarrow -t$ whereas E does not. Thus, if the minimum of the magnetic energy were to correspond to M pointing in a certain direction, the state with M pointing in the opposite direction would be another energy minimum with exactly the same energy. We, therefore, conclude that any magnetic state of the particle is at least doubly degenerate in the absence of an external magnetic field, Fig. 3.1. To make this statement more rigorous, one should take note of a special role played by nuclear spins [10,11]. The reversal of the magnetic moment of the particle means the reversal of all individual atomic moments. Imagine now that there are also non-zero nuclear spins in the particle. They are coupled with the atomic moments via hyperfine interaction. The latter depends on the mutual orientation of the nuclear and atomic spins. If the atomic spins were reversed and the nuclear spins were not, our argument about the double degeneracy of the state, based upon the time-reversal symmetry, would break. In other words, nuclear spins are equivalent to a weak non-zero external magnetic field. In order to preserve the degeneracy, one should assume that all spins, electronic and nuclear, are reversed simultaneously. This will be important for some but not all of the problems considered below. To move from simple to complex we will first neglect the effect of nuclear spins and will study it later.

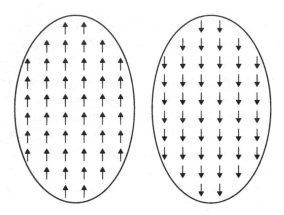

Figure 3.1 Degenerate states of a monodomain magnetic particle.

The degenerate states shown in Fig. 3.1 are separated by the energy barrier, U, due to the magnetic anisotropy. If the particle is in thermal equilibrium with the environment, the probability of the transition between these states is proportional to the Arrhénius factor $\exp(-U/T)$. The pre-exponential factor is of the order of the FMR frequency. At high temperature, thermal transitions are frequent and the particle behaves as a paramagnetic object of large magnetic moment. This effect is known as superparamagnetism. At low temperature, thermal transitions die out and, within the classical approximation, the magnetic moment of the particle freezes in the direction of one of the energy minima. Here we are interested in the possibility of quantum tunneling between the states shown in Fig. 3.1. Let us argue that, in order to exhibit such a tunneling, the particle must be firmly coupled with a solid matrix [29]. Indeed, for a free particle, the total angular momentum,

$$L + \gamma^{-1}M, \tag{3.3}$$

must be conserved. Here L is the angular momentum due to the mechanical rotation of the particle and $\gamma^{-1}M$ is its internal angular momentum due to the magnetic moment, γ being the gyromagnetic ratio. If M is of purely spin origin, $\gamma^{-1}M$ equals the total spin of the monodomain particle S. For specificity, we will have this situation in mind, unless specified otherwise. It is easy to see now that the transition $M \to -M$ cannot occur in a free non-rotating particle. Such a transition would cause a rotation in the final state with $L = 2\gamma^{-1}M$ and the kinetic energy, $L^2/(2I)$, associated with that rotation, I being the moment of inertia of the particle. This energy is not negligible. For a nanometer-size particle it can be of the order of a few kelvins, that is, much higher than any uncertainty in the energy of the particle at low temperature. Consequently, the energy of the final state would be significantly greater than the energy of the initial state, which forbids tunneling between M and $-M$. To compensate for the energy change, the particle should have been prepared in a metastable state by placing it in a sufficiently high magnetic field. If this is not the case, the particle could change the direction of its magnetic moment only if it were to have zero total angular momentum, that is, $L = -\gamma^{-1}M$. This follows from the fact that the energy of a free particle, in the absence of the field, is an even function both of M and of L and is invariant only under the simultaneous change of sign of these two vectors. Any rotation of M will then be accompanied by the rotation of L, such that $L + \gamma^{-1}M$ remains zero. If M is of spin origin this condition becomes $L + S = 0$. Since L is always an integer, it can be satisfied by an integer total spin S but not by a half-integer spin. Thus a half-integer spin cannot tunnel in the absence of a field under any circumstances. This is in accordance with Kramers' theorem in quantum mechanics that says that the double degeneracy of the magnetic state with respect to time reversal ($S \rightarrow -S$) cannot be removed for a half-integer spin. We will discuss this aspect of the problem in more detail later.

Although the above discussion shows that, for an integer S, tunneling between M and $-M$ is, in principle, possible for a rotating free particle, the experimental study of this effect is not feasible. We should, therefore, turn to a more practical and more typical situation in which a nanometer-size magnetic particle is embedded within a large solid matrix. The discussion of the free-particle case suggests that the possibility of tunneling should depend on how firmly the particle is coupled with the matrix. This is illustrated by Fig. 3.2. The intermediate state showing the elastic twist in the matrix should not be taken literally. It is a virtual state that occurs in imaginary time. In real time the particle goes directly from the state (a) to the state (c) in Fig. 3.2. Owing to the conservation of the total angular momentum, this transition, just like for a free particle, must generate mechanical rotation with the angular momentum $L = -2\gamma^{-1}M$. If this angular momentum is received by the entire matrix, the matrix will begin to rotate with the angular velocity L/I_m and the kinetic energy $L^2/(2I_m)$, where I_m is its moment of inertia. The latter is proportional to the fifth power of the size of the matrix, whereas L is proportional to the volume of the magnetic particle alone. Consequently, the corresponding angular velocity and kinetic energy already become negligible on any physical scale when the size of the matrix exceeds the

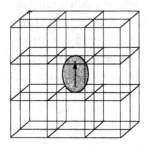 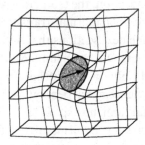 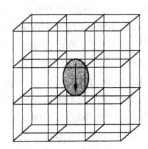

Figure 3.2 Tunneling of the magnetic moment of a single-domain particle embedded within an elastic matrix.

size of the particle by a factor of 100. A more careful analysis [29] shows that the angular momentum is transferred to the volume of dimension c_t/ω_0, where c_t is the speed of the transversal sound in the matrix and ω_0 is the instanton frequency. Typically this dimension is of the order of a few hundred nanometers, which is enough that one need not worry about the conservation of the angular momentum. However, in a soft matrix, with a small shear modulus, $\mu = \rho c_t^2$ (ρ being the mass density), the tunneling can be severely suppressed [29]. This makes the problem of spin tunneling different from the conventional double-well problem such as, e.g., tunneling between degenerate atomic configurations of the ammonia molecule. The latter tunneling trivially conserves the position of the center of mass of the molecule; it is not necessary for the molecule to be coupled with a large matrix. In the case of spin tunneling, however, the matrix is necessary in order to conserve the total angular momentum. The rotation of the matrix produces $2S$ transverse phonons of spin $s = 1$, each necessary to switch the total spin of the magnetic particle from S to $-S$. In the limit of an infinite absolutely rigid matrix, the angular velocity of this rotation tends to zero. So do the linear momenta of the transverse phonons. Thus, the absorption of the angular momentum by the matrix in the problem of spin tunneling is analogous to the absorption of the linear momentum by the matrix in the Mössbauer effect (the zero-phonon spectral line).

The $k = 0$ interaction of the crystalline lattice with the magnetic moment of the particle is due to the energy of the magnetic anisotropy. Transitions between different orientations of S can be also induced by placing the particle in a uniform magnetic field. In this case $k = 0$ photons of $s = 1$ are responsible for the conservation of the total angular momentum. For tunneling of a large spin, the number of these photons must be large. We, therefore, come to the conclusion that tunneling of large S, whether due to the anisotropy or due to the field, must always appear in the order of the perturbation theory that scales linearly with S. Thus, very generally, the leading term in the logarithm of the tunneling rate should be linear in the total tunneling spin, $\ln \Gamma_Q = \alpha - \beta S$. Notice that so should the logarithm of the thermal rate, $\ln \Gamma_T = \alpha' - U/T$, because the energy barrier U is proportional to the magnetic volume of the particle, $U \propto S$. As has been discussed in the previous section, the temperature T_c below which quantum transitions begin to dominate over thermal transitions can be estimated from $\beta S = U/T_c$. We shall see in the next sections that, in accordance with our qualitative ideas, for any kind of magnetic tunneling, S always cancels out from this equation, making T_c depend only on the intrinsic parameters of the magnetic material and the shape of the energy barrier, but not on the absolute height of the barrier and not on the total tunneling spin.

Let us now discuss possible mechanisms of the tunneling of the magnetic moment in monodomain particles. We have already mentioned the conditions of the uniform magnetization of a small particle, Eqs. (3.1) and (3.2). It is important to understand, however, that these conditions do not insure that the reversal of the magnetic moment of the particle occurs via uniform rotation of its magnetic moment. In fact, it has been demonstrated [26,30] that, for sufficiently large, *uniformly magnetized* particles, the magnetization reversal occurs via the *non-uniform curling mode*. The extreme case is a monodomain particle the size of which is large compared with the domain wall thickness. This becomes possible when the energy of the magnetocrystalline anisotropy is large compared with the magnetic dipole energy. In that case tunneling could, in principle, occur [31], due to the process shown in Fig. 3.3. Let us roughly estimate its probability. The WKB exponent for tunneling of a wall of mass m through a barrier of height U and width d has the order of magnitude $B \simeq (Umd^2)^{1/2}/\hbar$. Here U is due to the non-uniformity of the magnetization produced by the domain wall. It is roughly the energy of the wall in the middle of the particle, $U \simeq (\epsilon_{ex}\epsilon_{an})^{1/2}S^{2/3}$. The mass m can be estimated as the mass of 1 cm^2 of the wall (the Döring mass),

$$1/(\gamma^2\lambda) = \gamma^{-2}a^{-1}(\epsilon_{an}/\epsilon_{ex})^{1/2}, \tag{3.4}$$

times its mean area, $A \simeq d^2 \simeq S^{2/3}a^2$. To complete the estimate, one should take into account that $\frac{1}{2}\hbar\gamma$ equals the Bohr magneton μ_B and that the dipole energy per atom, ϵ_d, is of order μ_B^2/a^3. This gives $B \simeq (\epsilon_{an}/\epsilon_d)^{1/2}S$. Let us recall now that the condition $d \gg \lambda$, which is needed to make Fig. 3.3 meaningful, corresponds to the limit of a very large S. In addition, $\epsilon_{an} \gg \epsilon_d$. Consequently, the lifetime for the tunneling shown in Fig. 3.3 must be astronomically large, and this kind of magnetic tunneling should be unrealistic for monodomain particles. Note that the above consideration does not mean that domain walls cannot tunnel inside small particles. In our estimate of the energy barrier we assumed that there was no external magnetic field. The latter, if applied opposite to the magnetization of the particle, could significantly lower the energy barrier. Also, if the particle were full of defects, as many small particles are, the potential landscape for the domain wall could be rather complex. The wall can enter the particle and become trapped by the defects. It will then slowly move by thermal and quantum hoppings from one position to another, and may eventually exit the particle at a different location, reversing its magnetic moment. This process, as well as the field-induced entry of the wall into the particle, should not be very different

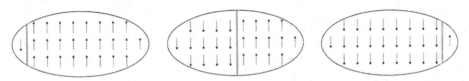

Figure 3.3 Magnetization reversal in a small particle by nucleation of a domain wall that sweeps across the particle.

from tunneling of domain walls in a bulk ferromagnet. The latter will be studied separately in Chapter 4 within a model that allows for the non-uniform rotation of the magnetization. When talking about monodomain particles, however, we will concentrate on particles whose size is smaller than the thickness of the domain wall, so that any non-uniformity in M is strongly suppressed by energy arguments. This is equivalent to the condition $S < (\epsilon_{ex}/\epsilon_{an})^{3/2}$. To be on the safe side, one can impose an even stronger condition, $S < \epsilon_{ex}/\epsilon_{an}$, which means that the energy required to rotate the magnetic moment of the particle as a whole is smaller than the exchange energy per atom.

3.2   Tunneling of the magnetic moment in monodomain ferromagnetic particles

The tunneling of the magnetic moment has certain similarities to the problem of the tunneling of a particle that we studied in Chapter 2. The instantons of the equation of motion for M, Eq. (1.1), and perturbations around them, determine the tunneling probability. There is, however, one essential difference from the conventional tunneling problem. The equation of motion for M, contrary to the equations of motion in mechanics, contains only the first-order derivative with respect to time. This equation describes the precession of the magnetic moment around a certain direction in space, determined by the magnetic anisotropy and the external magnetic field. The dissipation, due to the interaction of M with the lattice, conductivity electrons, nuclear spins, etc., eventually brings the magnetic moment to rest. For the time being we will neglect dissipation. Assuming, in accordance with the discussion of the previous section, that M has a fixed magnitude M, let us characterize this vector in terms of its spherical coordinates, θ and ϕ, Fig. 3.4. It is easy to show that, in terms of θ and ϕ, Eq. (1.1) is equivalent to the following two equations:

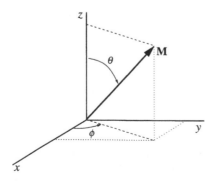

Figure 3.4   The spherical coordinate system used to describe the magnetic moment of a monodomain particle, M.

$$\frac{d\phi}{dt} = -\frac{\gamma}{M\sin\theta}\frac{\partial E}{\partial\theta}$$

$$\frac{d\theta}{dt} = +\frac{\gamma}{M\sin\theta}\frac{\partial E}{\partial\phi}, \tag{3.5}$$

where $E(\theta,\phi)$ is the magnetic energy of the particle.

We shall use Eq. (2.12) to compute the rate of tunneling of M out of a metastable state. For that we will need the path-integral definition of the partition function in the spirit of Eq. (2.26),

$$Z = \oint \mathcal{D}\{\theta\}\,\mathcal{D}\{\phi\}\,\exp\left(-\frac{1}{\hbar}\int_0^{\hbar/T} d\tau\,\mathcal{L}_E\right), \tag{3.6}$$

where $\mathcal{L}_E$ is the Euclidean magnetic Lagrangian. The latter is related to the real-time Lagrangian $\mathcal{L}$ through $\mathcal{L}_E = -\mathcal{L}(t \to -i\tau)$. To define $\mathcal{L}$ we need conjugate variables, $x(\tau)$ and $p(\tau)$. In the problem of a fixed-length angular momentum, $L = M/\gamma$, they are

$$x = \phi$$

$$p = \hbar L_z = \left(\frac{M}{\gamma}\right)\cos\theta. \tag{3.7}$$

The Lagrangian then is

$$\mathcal{L} = p\dot{x} - E(x,p) = \left(\frac{M}{\gamma}\right)\dot{\phi}\cos\theta - E(\theta,\phi) + \frac{df(\theta,\phi)}{dt}. \tag{3.8}$$

Notice that we have introduced into the Lagrangian the total time derivative of some function $f(\theta,\phi)$, which can always be added to a classical Lagrangian without changing the equations of motion. It is easy to see that the equations of motion that follow from (3.8),

$$\dot{x} = \frac{\partial E}{\partial p}$$

$$\dot{p} = -\frac{\partial E}{\partial x}, \tag{3.9}$$

coincide with Eqs. (3.5). We are now faced with the problem of choosing the function f in Eq. (3.8). The $\dot{f}$-term in the Lagrangian produces the phase in the exponent of Eq. (3.6), which is different for different $\theta(\tau),\phi(\tau)$ trajectories. Consequently, this phase term is important in the quantum problem. The correct choice is the one that gives a certain geometrical meaning to the magnetic action [32]. If one chooses

$$f = \frac{M}{\gamma}\phi, \tag{3.10}$$

the first term in the magnetic action,

$$I = \int dt \, \mathcal{L} = \int dt \left(\frac{M}{\gamma} \dot{\phi}(\cos\theta - 1) - E(\theta, \phi) \right), \tag{3.11}$$

is the area on the sphere of radius S swept by the total spin, $S = \gamma^{-1}M$, as it makes a closed path around the north pole of the sphere. The other name for that term is Berry's phase. Its significance for the problem of spin tunneling was elucidated by Loss et al. [33] and by von Delft and Henly [34].

We will compute the tunneling exponent (the Euclidean action) for the four typical situations, which we will call models I–IV.

3.2.1   Model I

Our model I is a doubly degenerate state of a particle of magnetic volume V and of total anisotropy energy

$$E = k_\perp M_z^2 - k_\parallel M_x^2 = V(K_\perp \cos^2\theta - K_\parallel \sin^2\theta \cos^2\phi). \tag{3.12}$$

In the second part of this formula we have redefined the constants to explicitly show the proportionality of the energy to V. At $K_\perp > 0$ and $K_\parallel > 0$ it describes the X–Y easy plane and the X easy direction in that plane. The energy (3.12) has two degenerate minima corresponding to M looking in the positive and negative X directions, that is, $\phi = 0, \pi$ at $\theta = \pi/2$. The barrier between these two states has a minimum value, $U = K_\parallel V$, when the rotation of M occurs in the X–Y plane. The maximum value of the barrier, $U = K_\perp + K_\parallel$, corresponds to the rotation in the X–Z plane.

We need to compute the integral

$$\oint \mathcal{D}\{\cos\theta\} \mathcal{D}\{\phi\} \exp \left[-\frac{V}{\hbar} \int d\tau \right.$$
$$\left. \times \left(i\frac{M_0}{\gamma}\dot{\phi}_\tau - i\frac{M_0}{\gamma}\dot{\phi}_\tau \cos\theta + (K_\perp + K_\parallel \cos^2\phi)\cos^2\theta + K_\parallel \sin^2\phi \right) \right], \tag{3.13}$$

where $\dot{\phi}_\tau$ means the derivative with respect to τ. Note that, to make the exponent zero for the equilibrium states (M parallel to X), we have added the height of the barrier, $K_\parallel V$, to the Lagrangian.

Let us first integrate over the $\theta(\tau)$ trajectories which begin and end at $\theta = \pi/2$. The integral over $\cos\theta$ is Gaussian and can be performed explicitly. This leaves one with the following integral over $\phi(\tau)$ trajectories:

$$\oint \mathcal{D}\{\phi\} \exp \left[-\frac{V}{\hbar} \int d\tau \left(i\frac{M_0}{\gamma}\dot{\phi} + \frac{M_0^2 \dot{\phi}^2}{4\gamma^2(K_\perp + K_\parallel \cos\phi^2)} + K_\parallel \sin\phi^2 \right) \right]. \tag{3.14}$$

There are two extremal trajectories that minimize the Euclidian action in the exponent of this expression, at $T = 0$ [4],

$$\phi = \pm \arccos \left(-\frac{K_\perp^{1/2} \sinh (\omega_0 \tau)}{[K_\parallel + K_\perp \cosh^2 (\omega_0 \tau)]^{1/2}} \right), \tag{3.15}$$

where we have introduced

$$\omega_0 = \frac{2\gamma}{M_0} [K_\parallel (K_\parallel + K_\perp)]^{1/2}. \tag{3.16}$$

These solutions carry out the clockwise and counterclockwise underbarrier rotations of the magnetic moment from $\phi = 0$ at $\tau = -\infty$ to $\phi = \pm \pi$ at $\tau = +\infty$. Note that they exist only for $K_\perp \neq 0$; that is, when M_x does not commute with the Hamiltonian. The contribution of the instantons (3.15) to the path integral determines the WKB exponent of the tunneling rate. Small perturbations around instantons give the pre-exponential factor. Analysis similar to that of Chapter 2 shows that, on the order of magnitude, the pre-exponential factor, A, for spin tunneling is always the instanton frequency times $(B/2\pi)^{1/2}$. The latter factor can hardly be greater than 2–3; if B is greater than 30, tunneling would be difficult to observe. Thus, the instanton frequency provides a good estimate of A. Typically it is in the range 10^9–10^{11} s^{-1}. The computation of the exact value of the pre-factor, albeit of a certain esthetic value, in most cases is not worth the effort since the dependence of the tunneling rate on the control parameters (temperature, field, volume, orientation, etc.) is dominated by the exponent.

For each trajectory, the term in the action which is linear in $\dot{\phi}$ generates the phase factor $\exp (-iS \Delta \phi)$, where $\Delta \phi$ is the total change in ϕ along the trajectory. Here we have introduced the total spin of the particle, $S = M_0 V/(\hbar \gamma)$. It is now easy to see that phases generated by clockwise ($\Delta \phi = -\pi$) and counterclockwise ($\Delta \phi = \pi$) instantons combine into $\cos (S\pi)$, which is ± 1 or 0 depending on whether S is an integer or a half-integer. This implies that, in the absence of the magnetic field, an integer S can tunnel but a half-integer S cannot. Note that tunneling removes the degeneracy of the ground state with respect to the orientation of M. According to Kramers' theorem, this may occur only for an integer S. Our path integral method for the spin is, therefore, in agreement with conventional quantum mechanics [33,34].

If a magnetic field is applied along the anisotropy axis, the Kramers degeneracy is removed, and tunneling is no longer prohibited for a half-integer spin. It is clear, however, from the expected non-singular dependence of Γ on H that, in the presence of a small field, tunneling must still be suppressed for a half-integer S compared with the tunneling rate for an integer S. One may ask, therefore, how small the field should be for one to observe the difference in tunneling rates for an integer and a half-integer spin. Analysis on the basis of the coherent spin-state representation [35] shows that the freezing of the half-integer S should occur in a field $H < H_\parallel /S$, where $H_\parallel = 2K_\parallel /M_0$ is the longitudinal anisotropy field. For a nanometer-size particle with $H_\parallel \simeq 10^4$ Oe and $S \simeq 10^4$ this gives about 1 Oe. In an experiment with a large number of single-domain particles of different sizes one should expect statistically equal numbers of integer and half-integer spins. If all moments of the particles are initially magnetized in one direction and then the field is switched off, the freezing effect should reveal itself in a longer magnetic relaxation for the half-integer particles. This should result in a peculiar time dependence of the relaxation: a fast drop of the magnetic moment of the system to one-half of the initial value and then slow relaxation to zero. Of course, to observe this effect one must still have a very narrow distribution of particle sizes, otherwise the broad distribution of individual lifetimes will smear relaxation.

To within a numerical factor, the result for the tunneling rate reads [4]

$$\Gamma \simeq |\cos(S\pi)|\, \omega_0 \exp\left\{-2S\, \ln\left[\left(1 + \frac{K_\parallel}{K_\perp}\right)^{1/2} + \left(\frac{K_\parallel}{K_\perp}\right)^{1/2}\right]\right\}. \qquad (3.17)$$

For an integer S it gives

$$\Gamma \simeq \omega_0 \exp\left[-S\, \ln\left(\frac{4K_\parallel}{K_\perp}\right)\right] = \omega_0 \left(\frac{K_\perp}{4K_\parallel}\right)^S \qquad (3.18)$$

at $K_\perp \ll K_\parallel$, and

$$\Gamma \simeq \omega_0 \exp\left[-2S\left(\frac{K_\parallel}{K_\perp}\right)^{1/2}\right] \qquad (3.19)$$

at $K_\perp \gg K_\parallel$. The latter case is of major interest for tunneling of a large spin. It may apply to Tb and Dy, for which $K_\parallel \simeq 10^6$ erg cm^{-3} and $K_\perp \simeq 10^8$ erg cm^{-3}. These would give $\omega_0 \simeq 10^{11}$ s^{-1} and the tunneling rate $\Gamma \simeq 10^{11}$ exp $(-0.2\,S)$. Correspondingly, even in the absence of a field, it may still be possible to observe tunneling of S as large as 200. In the opposite case of almost uniaxial anisotropy it is important to notice that, although the WKB exponent becomes infinite in the limit of $K_\perp = 0$, it increases only logarithmically as $K_\perp$ goes to zero. That is, the tunneling rate decreases as a power of $K_\perp$. For that reason, tunneling of a not very large spin may be possible even in this case.

The rate of quantum tunneling should be compared with the rate of thermal transitions, which goes as exp $(-K_\parallel V/T)$. One finds that quantum transitions dominate below

$$T_c = \mu_B H_\parallel / \ln\,(4H_\parallel/H_\perp) \qquad (3.20)$$

for $K_\perp \ll K_\parallel$, and below

$$T_c = \tfrac{1}{2}\mu_B (H_\parallel H_\perp)^{1/2} \qquad (3.21)$$

for $K_\perp \gg K_\parallel$. Here we have introduced anisotropy fields, $H_{\parallel(\perp)} = 2K_{\parallel(\perp)}/M_0$, and the Bohr magneton, $\mu_B = \hbar\gamma/2$. For $H_\parallel \simeq 10^4$ Oe and not very small $H_\perp$ this gives T_c of the order of 1 K in accordance with values reported from magnetic relaxation measurements for many materials (see Chapter 5). The tunneling rate can significantly increase in the presence of a magnetic field that lowers the barrier. These is illustrated by the next three models considered below.

3.2.2   Model II

Our next example (Fig. 3.5) is a uniaxial single-domain particle, with the anisotropy axis the Z-axis, in a transverse magnetic field applied along the X-axis:

$$E = V\left(K\sin^2\theta - HM_0\sin\theta\cos\phi + \frac{H^2 M_0^2}{4K}\right). \tag{3.22}$$

For $H < H_c = 2K/M_0$, the energy (3.22) has two minima ($E_{\min} = 0$). They correspond to the two states shown in Fig. 3.5 with $\theta = \theta_0$ and $\theta = \pi - \theta_0$ at $\phi = 0$, where $\theta_0 = H/H_c$. The energy barrier between the two states is $U = K\epsilon^2$, where $\epsilon = 1 - H/H_c$. At $H = H_c$, when $\theta_0 = \pi/2$, the two equilibrium orientations of M meet on the X-axis and the barrier disappears.

Equations (3.5) with the energy (3.22) produce the following equation for θ in imaginary time [4]:

$$\frac{d^2\theta}{d\tau^2} = \omega_H^2\left[1 + \omega_H^{-2}\left(\frac{d\theta}{d\tau}\right)^2\right]\cot\theta - 2\omega_1\omega_H\left[1 + \omega_H^{-2}\left(\frac{d\theta}{d\tau}\right)^2\right]^{1/2}\cos\theta,$$

$$\tag{3.23}$$

where $\omega_H = \gamma H$ and $\omega_1 = \gamma K/M_0$.

Consider first the limiting case of a very low field, $H \to 0$, when $\theta_0 \to 0$. In this case the two states shown in Fig. 3.5 become very close to M parallel and antiparallel to the Z-axis. They are switched by the instanton of Eq. (3.23) at $H \to 0$:

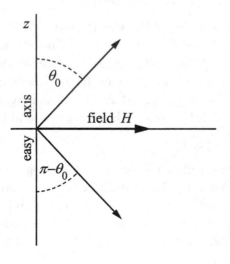

Figure 3.5 Two classically degenerate magnetic states for model II.

$$\theta = \arccos\left[\tanh\left(\omega_1 \tau\right)\right]. \tag{3.24}$$

The calculation of the tunneling exponent via Eq. (3.11) then gives [4,8,9]

$$B = 2S \ln\left(\frac{H_c}{H}\right). \tag{3.25}$$

Note that this answer can, in fact, be anticipated from the conventional approach via the Schrödinger equation. Indeed, it means that, in the limit of small H, the tunneling rate is proportional to $(H/H_c)^{2S}$. This is easy to see from the fact that the transition between $S_z = S$ and $S_z = -S$ appears in the $2S$ order of the perturbation theory on the transverse field H. The exact expression for the rate in that limit will be given in Chapter 7, in connection with the problem of spin tunneling in Mn-12 molecules.

Consider now the opposite limit of $H \to H_c$, that is $\epsilon \to 0$. By introducing a small angle $\delta = \pi/2 - \theta$ between M and the X-axis, one obtains from Eq. (3.23) the instanton,

$$\delta = (2\epsilon)^{1/2} \tanh\left[(\epsilon)^{1/2}\omega_H \tau\right], \tag{3.26}$$

which switches M between the two energy minima, $\delta = \pm(2\epsilon)^{1/2}$. In this case [4]

$$B = 4S\epsilon^{3/2}. \tag{3.27}$$

3.2.3   Model III

Our next model has the same structure of the magnetic anisotropy as model I but, in addition, it has the magnetic field applied in the negative X direction, opposite to the magnetic moment of the particle. The magnetic energy of the particle then becomes

$$E = k_\perp M_z^2 - k_\parallel M_x^2 + M_x H = V(K_\perp \cos^2 \theta - K_\parallel \sin^2 \theta \cos^2 \phi \\ + M_0 H \sin \theta \cos \phi). \tag{3.28}$$

Now the two energy minima are inequivalent, M looking in the positive X direction being a metastable state and M looking in the negative X direction being the absolute minimum of the energy. The energy barrier between the two minima is $K_\parallel V \epsilon^2$, where $\epsilon = 1 - H/H_\parallel$. We shall assume the most favorable situation of $K_\perp \gg K_\parallel$ and $H \to H_\parallel$. In this case, the angle that M should rotate to reach the other side of the barrier is $\phi = 2\sqrt{\epsilon} \ll 1$. For small ϕ the energy can be written as

$$E = V[K_\perp \cos^2 \theta + K_\parallel(\epsilon\phi^2 - \tfrac{1}{4}\phi^4)]. \tag{3.29}$$

On recalling that $\cos \theta$ is our generalized momentum and ϕ is our generalized coordinate, we see that Eq. (3.29) and the Lagrangian in Eq. (3.11) are now equivalent to the problem of a non-relativistic particle, Eqs. (2.21) and (2.22). Note that, although in general the magnetic Hamiltonian does not have the form (2.23), for this particular problem it reduces to that form. The answer for the tunneling rate can be obtained by using the same method as for the previous model or the conventional WKB method. It reads [4]

$$\Gamma \simeq \omega_H a(H) \exp\left[-\frac{8}{3} S \left(\frac{K_\parallel}{K_\perp}\right)^{1/2} \epsilon^{3/2}\right],$$ (3.30)

where $a(H)$ is a dimensionless function that oscillates in the magnetic field [35], and

$$\omega_H = \frac{2\gamma}{M_0}(\epsilon K_\parallel K_\perp)^{1/2}.$$ (3.31)

For $M_0 \simeq 10^2$ emu, $K_\parallel \simeq 10^6$ erg cm^{-3}, $K_\perp \simeq 10^8$ erg cm^{-3}, and $\epsilon \simeq 10^{-2}$, we obtain $\Gamma \simeq 10^{11} \exp(-\frac{8}{3} \times 10^{-4} S)$ s^{-1}.

Thus, *by applying the magnetic field that lowers the energy barrier, one can, in principle, obtain a significant tunneling rate for a total spin as large as 10^5*, that is, for a particle of considerable size. Note that, for the above numbers, $H_\parallel \simeq 10^4$ Oe, which means that $\epsilon \simeq 10^{-2}$ can be easily achieved by controlling the field to within an accuracy of only 100 Oe. The fine tuning of the field to the value of $H_\parallel$ will provide much lower barriers and the possibility of tunneling of an even greater total spin of the particle. One should remember, however, that a small ϵ means a small tunneling angle. To talk about macroscopic quantum tunneling, the magnetic states on the two sides of the barrier should be macroscopically different. For $\epsilon \simeq 10^{-2}$, the tunneling angle for the magnetic moment is $\phi = 2\sqrt{\epsilon} = 0.2$ rad$= 11.5°$, that is, quite large enough for an experimentalist to distinguish easily between the two states. Model III and the next model IV have been the basis for recent experiments on individual single-domain particles discussed in Section 6.3.

3.3   The angular dependence of the tunneling rate (model IV)

In models II and III we assumed that the field was applied at a right angle or parallel to the magnetic moment. The generic problem, however, and the easiest to implement in practice, is that of uniaxial anisotropy with the magnetic field applied at a some angle θ_H to the anisotropy axis [36,37]. This problem does not possess any symmetry and for that reason is more difficult mathematically. It is worth pursuing, however, because of its significance for experiments. Consider a uniaxial single-domain particle whose anisotropy direction coincides with the z-axis. In the absence of the field there are two equilibrium orientations of M: along and against the z-direction. Let us now assume that the magnetic field is applied in the Z–X plane, at an angle $90° \leq \theta_H \leq 180°$ to the z-axis. If the field is below some critical value, $H_c(\theta_H)$ (to be computed), M has the two equilibrium orientations shown in Fig. 3.6. One of them, with $0° \le \theta \le 90°$, is a metastable state; the other, with $90° \le \theta \le 180°$, corresponds to the absolute minimum of the magnetic energy. We are interested in the quantum decay of the metastable state.

The magnetic energy of the particle is the sum of its anisotropy energy and its Zeeman energy,

$$E = -kM_z^2 - M_x H_x - M_z H_z, \tag{3.32}$$

where $k > 0$ is a dimensionless anisotropy constant. For an arbitrary orientation of M the energy can be rewritten as

$$E = -H_a M (\tfrac{1}{2}\cos^2\theta + h_z \cos\theta + h_x \sin\theta\cos\phi), \tag{3.33}$$

where we have introduced the anisotropy field, $H_a = 2kM$, and dimensionless components of the magnetic field, $h_{x,z} = H_{x,z}/H_a$; θ and ϕ are the conventional spherical coordinates of the fixed-length vector M. The metastable state exists at $H_z < 0$. It corresponds to M in the X–Z plane (that is, $\phi = 0$) at some angle $\theta = \theta_0$. Near this point the potential has the form of a canyon with its bottom at $\phi = 0$ satisfying $E = (H_a M)E_\theta$, where

$$E_\theta = -\tfrac{1}{2}\cos^2\theta - h\cos(\theta - \theta_H), \tag{3.34}$$

Here we have switched to θ_H and $h = H/H_a$ from $h_x = h\sin\theta_H$ and $h_z = h\cos\theta_H$. The dependence of E_θ on θ, near θ_0, at various values of the field, is schematically shown in Fig. 3.7.

The metastable state satisfies

$$\sin(2\theta_0) + 2h\sin(\theta_0 - \theta_H) = 0, \tag{3.35}$$

which follows from $\partial E_\theta / \partial\theta = 0$. The point $\theta = \theta_1$ is the saddle point of the potential $E(\theta, \phi)$. For the metastable state to decay, M must either fluctuate to the top of the barrier at $\theta = \theta_1$ or tunnel under the barrier to the escape point $\theta = \theta_2$. Since M is considered to be a macroscopic variable, an appreciable escape rate is expected only when the barrier is lowered by tuning the external

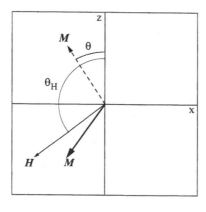

Figure 3.6 The geometry of the problem. Vectors M shown by dashed and solid lines correspond to the metastable and stable states, respectively.

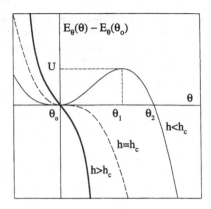

Figure 3.7 The θ-dependence of the potential for different values of the field.

field to the critical field $h_c(\theta_H)$. At $h = h_c$ all three angles in Fig. 3.7 coincide, $\theta_o = \theta_1 = \theta_2 = \theta_c$, and both the first and the second derivative of E_θ become zero at $\theta = \theta_c$. This gives two equations for θ_c and h_c:

$$\sin(2\theta_c) + 2h_c \sin(\theta_c - \theta_H) = 0$$
$$\cos(2\theta_c) + 2h_c \cos(\theta_c - \theta_H) = 0. \tag{3.36}$$

On solving them we obtain

$$\sin^3 \theta_c = h_c \sin \theta_H$$
$$\cos^3 \theta_c = -h_c \cos \theta_H \tag{3.37}$$

$$h_c = (\sin^{2/3} \theta_H + |\cos \theta_H|^{2/3})^{-3/2}. \tag{3.38}$$

Note that all signs in Eq. (3.37) are in accordance with the fact that $0° \le \theta_c \le 90°$ while $90° \le \theta_H \le 180°$. The dependence of h_c on θ_H is plotted in Fig. 3.8.

Let us now consider a field that is slightly lower than the critical field,

$$h = h_c(1 - \epsilon), \tag{3.39}$$

where $\epsilon \ll 1$. Such a field still preserves the metastable state. The corresponding equilibrium value of θ is now slightly lower than θ_c, $\theta_c - \theta_o = \Delta \ll 1$. By expanding Eq. (3.35) near θ_c we obtain with the help of Eqs. (3.37) $\Delta = (2\epsilon/3)^{1/2}$. Then introducing a positive $\delta = (\theta - \theta_o) \ll 1$, and expanding Eq. (3.34) near θ_o, we get, to the third order in δ,

$$E_\theta(\theta) - E_\theta(\theta_o) = \tfrac{1}{4}\sin(2\theta_c) \left[(6\epsilon)^{1/2}\delta^2 - \delta^3\right]. \tag{3.40}$$

The escape point is $\delta_2 = \theta_2 - \theta_o = (6\epsilon)^{1/2}$.

The imaginary-time action for this problem can be written as

$$I = -\hbar S \int d\bar\tau \left(i\dot\phi_{\bar\tau}(\cos\theta - 1) + \tfrac{1}{2}\cos^2\theta + h_z\cos\theta + h_x\sin\theta\cos\phi\right),$$

$$\tag{3.41}$$

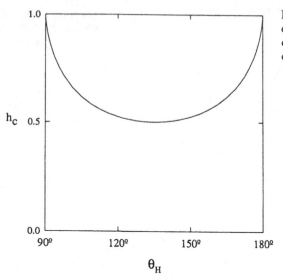

Figure 3.8 The dependence of the critical value of the field on the field orientation.

where we have introduced $\omega_a = \gamma H_a$, the dimensionless imaginary time $\bar{\tau} = \omega_a \tau$, and the total spin of the particle, $S = M/(\hbar\gamma) \gg 1$. To obtain the tunneling exponent, one must consider the extremal trajectories of Eq. (3.41):

$$i\dot{\theta}_{\bar{\tau}} = h_x \sin\phi$$

$$i\dot{\phi}_{\bar{\tau}} \sin\theta = h_x \cos\theta \cos\phi - h_z \sin\theta - \cos\theta \sin\theta. \tag{3.42}$$

Note that, as usual, the generalized coordinate and momentum can be interchanged in our calculation. Instead of considering real ϕ and imaginary $\cos\theta$, one can consider imaginary ϕ and real θ. In fact, the latter is the only appropriate choice because, according to the chosen geometry (see Fig. 3.6), θ is the real tunneling coordinate while $\phi = 0$ both in the initial and in the final state.

Equations (3.42) have an instanton solution that carries out the underbarrier rotation of M to the escape point. For this solution both θ and ϕ depend on τ. It starts at $\theta = \theta_0$, $\phi = 0$ at $\tau = -\infty$, comes to $\theta = \theta_2$, $\phi = 0$ at $\tau = 0$, and then bounces back to $\theta = \theta_0$, $\phi = 0$ at $\tau = +\infty$. The general solution is difficult to obtain. We shall recall that we are interested in the case of a low barrier when $h \to h_c$. In this case the potential in Fig. 3.7 becomes nearly flat and the $\bar{\tau}$ derivative of θ must be proportional to some power of a small parameter ϵ. This is the approximation of a *slow* instanton, which is easy to understand if one notices that the instanton corresponds to the classical motion in the inverted potential (which is now almost flat). According to the first of Eqs. (3.42), this means that the instanton involves only small ϕ; that is, the classical trajectory lies near the bottom of the potential *canyon*, close to the X–Z plane. The phase term in Eq. (3.41), proportional to $\dot{\phi}_{\bar{\tau}}$ (not the $\dot{\phi}_{\bar{\tau}}\cos\theta$ term), is important for tunneling between equivalent minima [33,34]. For the closed instanton trajectory described above it makes no contribution to the integral and, therefore, can be omitted.

The tunneling exponent B follows from the path integral

$$\int \mathcal{D}\{\phi(\tau)\} \int \mathcal{D}\{\cos\theta(\tau)\} \, \exp\left(-\frac{I}{\hbar}\right) \qquad (3.43)$$

over the continuum of trajectories that start and end at the metastable state $\theta = \theta_o$, $\phi = 0$ and that are close to the instanton. After integrating $\dot{\phi}_{\bar{\tau}} \cos\theta$ in Eq. (3.41) by parts, taking account of the boundary condition $\phi(\pm\infty) = 0$, and using the smallness of ϕ along the instanton, the path integral that we have to compute becomes

$$\int \mathcal{D}\{\phi(\tau)\} \int \mathcal{D}\{\cos\theta(\tau)\}$$
$$\times \exp\left[-S \int_{-\infty}^{\infty} \mathrm{d}\bar{\tau} \left(-i\dot{\theta}_{\bar{\tau}}(\phi\sin\theta) + \frac{h\sin\theta_H}{2\sin\theta}(\phi\sin\theta)^2 + E_\theta(\theta)\right)\right],$$
$$(3.44)$$

where $E_\theta(\theta)$ is given by Eq. (3.34). One can now select the new variables of the functional integration, $\phi\sin\theta$ and $\cos\theta$, and notice that the integration over $\phi\sin\theta$ is Gaussian. This gives

$$B = S \int_{-\infty}^{\infty} \mathrm{d}\bar{\tau} \left(\frac{\dot{\theta}_{\bar{\tau}}^2 \sin\theta_c}{2h_c \sin\theta_H} + E_\theta(\theta) - E_\theta(\theta_o)\right). \qquad (3.45)$$

Note that we put $\theta = \theta_c$ and $h = h_c$ in the first term because $\dot{\theta}_{\bar{\tau}}^2$ already has a smallness of order $\sqrt{\epsilon}$ coming from the $\bar{\tau}$ derivative, so that this term is proportional to $\epsilon^{3/2}$ and is of the same order in ϵ as are the other terms in Eq. (3.45) (this claim is justified below). A constant, $E_\theta(\theta_o)$, is added to make $B = 0$ at $\theta = \theta_o$. The integration with respect to $\bar{\tau}$ must now be performed over the instanton $\theta(\bar{\tau})$ that minimizes B. This simplifies the problem tremendously, compared with the problem in which the action depended on $\phi(\bar{\tau})$ and $\theta(\bar{\tau})$, though a complete mathematical equivalence to the initial problem is preserved. We shall now use the fact that, in the limit of a small barrier the instanton involves small deviations from θ_o, and replace θ in Eq. (3.45) by $\theta_o + \delta$. With the help of Eq. (3.40) we obtain

$$B = S \int_{-\infty}^{\infty} \mathrm{d}\bar{\tau} \left(\frac{\dot{\delta}_{\bar{\tau}}^2 \sin\theta_c}{2h_c \sin\theta_H} + \frac{1}{4}\sin(2\theta_c)\left[(6\epsilon)^{1/2}\delta^2 - \delta^3\right]\right). \qquad (3.46)$$

The extremal trajectory of Eq. (3.46) is

$$\delta(\tau) = \frac{\delta_2}{\cosh^2(\omega_o\tau/2)}, \qquad (3.47)$$

where

$$\omega_o = (6\epsilon)^{1/4} \frac{|\cot \theta_H|^{1/6}}{1 + |\cot \theta_H|^{2/3}} \omega_a. \tag{3.48}$$

Expressions (3.37) and (3.38) have been used to obtain the dependence of ω_o on θ_H. As required, the instanton starts at $\delta = 0$ ($\theta = \theta_o$) at $\tau = -\infty$, comes to the escape point $\delta = \delta_2$ ($\theta = \theta_2$) at $\tau = 0$, and then bounces back to $\delta = 0$ ($\theta = \theta_o$) at $\tau = +\infty$. Substitution of Eq. (3.47) into Eq. (3.46) gives, with the help of Eqs. (3.37) and (3.38), a simple formula for B [37], namely

$$B = \frac{16 \times 6^{1/4}}{5} S\epsilon^{5/4}|\cot \theta_H|^{1/6}. \tag{3.49}$$

Note that this answer also follows from a very different approach [36] in which the spin Hamiltonian is mapped onto a particle Hamiltonian that has the same structure of low-lying energy levels [38].

The angular dependence of B is plotted in Fig. 3.9. We see that B has a rather unexpected distinct dependence on the orientation of the field. It is almost flat for θ_H not close to 90° or 180°, rises sharply as θ_H approaches 180° in accordance with the fact that, at $\theta_H = 180°$, M_z commutes with the Hamiltonian, and rapidly drops at $\theta_H \to 90°$. The latter result must be taken with caution because, at $\theta_H = 90°$, the problem possesses a symmetry that has not been taken explicitly into account. In that limit it becomes our model II for the tunneling between equivalent minima. The potential then acquires the form $\delta^2 - \delta^4$ and the answer for B reads $B = 4S\epsilon^{3/2}$. Consequently, the result (3.49) holds for $\theta_H = 90° + \beta$, where β is large compared with $\epsilon^{3/2}$. This means that for, e.g., $\epsilon = 10^{-2}$, Eq. (3.49) is certainly true outside the 1° vicinity of 90°.

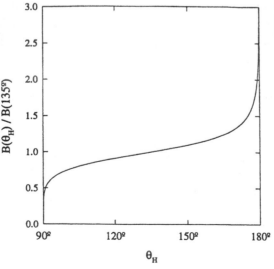

Figure 3.9 The angular dependence of the tunneling exponent.

Of course, one should remember that all our conclusions on the angular and field dependences of the tunneling exponent are valid in the limit of a low barrier. For a particle of considerable size this should be the only limit accessible to experiment. It is important, therefore, to analyze the validity of the semiclassical approximation. Obviously, for the method to be valid, the tunneling probability must be small; that is, $B \gg 1$ is required. This may not be enough, however. One should also worry that the energy $\hbar\omega$ of zero-point oscillations at the bottom of the metastable well is sufficiently small compared with the height of the barrier, U. It follows from Eq. (3.46) that the zero-point energy is $\hbar\omega_0$. The barrier height can be computed from Eq. (3.40),

$$U = \hbar\omega_a S \left(\frac{2\epsilon}{3}\right)^{3/2} \sin\left(2\theta_c\right). \tag{3.50}$$

By working out the ratio, we obtain

$$\frac{U}{\hbar\omega_0} = \frac{5}{36} B. \tag{3.51}$$

The optimal range of B in experiments on small particles should be $B \simeq 30\text{--}35$ (see below). In this case the semiclassical approximation should be rather good.

At a sufficiently large temperature, the tunneling exponent must cross to the Boltzmann exponent U/T. Then, by equating B of Eq. (3.49) to U/T, we obtain that the crossover from the quantum to the thermal regime occurs at

$$T_c = \frac{5}{36} \hbar\omega_0, \tag{3.52}$$

where we have used Eqs. (3.37) and (3.38). Another method to obtain the crossover temperature (see Chapter 2) is based upon the analysis of $\phi(\tau)$, $\theta(\tau)$ solutions of Eq. (3.42) that are periodic in τ with the period $\hbar/T$. At a finite temperature only such trajectories contribute to the path integral. Equations (3.42) for a class of periodic trajectories have a bifurcation at $T'_c = \hbar\omega/(2\pi)$, where ω is the frequency of small oscillations near the bottom of the inverted potential in Fig. 3.7. Above this temperature the only trajectory of interest which is formally periodic with the period $\hbar/T$ is a static solution of Eq. (3.42); $\theta = \theta_1$, $\phi = 0$. It corresponds to M at the top of the potential barrier. Substitution of this solution into Eq. (3.41) gives $B = U/T$. A simple analysis yields $\omega = \omega_0$, that is,

$$T'_c = \frac{\hbar\omega_0}{2\pi}. \tag{3.53}$$

Comparison of Eq. (3.52) with Eq. (3.53) shows that they differ by a factor of about 1.15, and, therefore, can both be used as the definition of the crossover temperature.

Because of the exponential dependence of the thermal rate on the temperature, the transition from the thermal to the quantum regime must become rather sharp as the temperature is lowered, with a well-defined crossover. The dependence of T_c on θ_H is plotted in Fig. 3.10. The plot suggests that the observation of the quantum decay of a metastable state at $\theta_H \to 90°$ and $\theta_H \to 180°$ requires much lower temperatures than are required for intermediate orientations of the field. Again, the vicinity of 90° must be treated with caution. In that region T_c does not, in fact, go to zero but is proportional to a higher power of the small parameter ϵ [4], $\epsilon^{1/2}$ instead of $\epsilon^{1/4}$ for intermediate angles. On writing θ_H as $90° + \beta$ we again find that our result for T_c is correct for $\beta > \epsilon^{3/2}$, that is, quite close to 90° for $\epsilon \le 10^{-2}$.

The geometry of model IV can be most advantageous for tunneling studies. It gives the experimentalist three control parameters for comparison with the theory: the orientation of the field, θ_H; the field strength $\epsilon = 1 - H/H_c$; and the temperature T. According to Eqs. (3.48), (3.52), and (3.53), the crossover temperature is proportional to the anisotropy field. Thus, selecting a particle with $H_a \ge 1$ T would insure that a reasonably low temperature is required. The weak dependence of T_c on ϵ, $T_c \propto \epsilon^{1/4}$, works to the advantage of an experimentalist. For, e.g., $H_a = 1$ T and $\epsilon = 10^{-3}$ we obtain $T_c(135°) \simeq 30$ mK. Note that $\epsilon \simeq 10^{-3}$ requires that the field be controlled to within an accuracy higher than $H_a \epsilon$, that is, to within better than 10 Oe for $H_a \simeq 1$ T.

If the experiment is to be performed with a large number of identical particles, the long lifetime of a metastable state will not necessarily interfere with the possibility of detecting the effect (compare with radioactive decay). In the case of a single particle, however, a very long, as well as a very short, lifetime can be a serious obstacle, so that special consideration should be given to the selection of the appropriate size of the particle. For $H_a \simeq 1$ T the prefactor of the tunneling

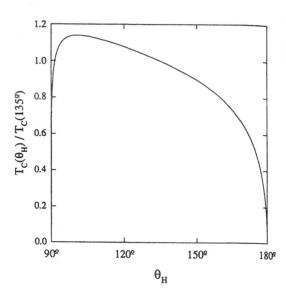

Figure 3.10 The dependence of the crossover temperature on the orientation of the magnetic field.

rate is of the order of 10^{11} s^{-1}. Consequently, a tunneling exponent significantly less than 30 will make tunneling very fast and, therefore, difficult to observe. It will also invalidate the semiclassical approximation. On the other hand, for B greater than 30–35, the lifetime of the metastable state can significantly exceed the time of the experiment. To insure that the lifetime of the metastable state does not exceed the time of the experiment, $B \simeq 30$ must be used for the observation of tunneling and, at the same time, for remaining within the domain of MQT. According to Eq. (3.49), this means that, at $\epsilon \simeq 10^{-3}$, the total spin of the particle S should be no more than 3×10^4. This certainly falls within the domain of MQT. Note that, even for ϵ as small as 10^{-3}, the underbarrier rotation of M corresponds to an appreciable change in the orientation, $\delta_2 = (6\epsilon)^{1/2}$ rad $> 4°$. Greater ϵ will give a larger angle and a higher T_c but will move the maximal allowed S to lower values. Smaller ϵ are dangerous because of the interaction of M with nuclear spins (see below).

The following experimental procedure can be applied. First, the anisotropy axis of the particle must be precisely determined by, e.g., allowing it to orient itself freely in the magnetic field. The position of the particle must then be fixed at a certain angle θ_H to the field, see Fig. 3.6. The magnitude of H should then be slowly increased to obtain the critical field, $H_c(\theta_H)$, at which the barrier disappears. The metastable state should then be created again and the field should be tuned to the value $H_c(1 - \epsilon)$ just below H_c. By repeating this procedure many times the transition rate must be obtained. The field orientation then should be changed and the whole procedure repeated for different values of θ_H and $H_c(\theta_H)$ but with the same value of ϵ. This would give the transition rate as a function of θ_H, which can be compared with the theoretical prediction given by Eq. (3.49) and illustrated in Fig. 3.9. To insure that quantum, rather than thermal, transitions are measured, the temperature independence of the rate should be checked at any θ_H.

3.4 Tunneling of the Néel vector in antiferromagnetic particles

Similar, even stronger, quantum effects exist in antiferromagnetic particles in which tunneling of the Néel vector l must result in a quantum superposition of antiferromagnetic sublattices, Fig. 3.11. Magnetic tunneling in ferromagnets is due to the terms in the magnetic anisotropy which violate commutation of M with the Hamiltonian. It is, therefore, a relativistic effect. In antiferromagnets the well-known exchange enhancement of the anisotropy makes tunneling of the Néel vector a much stronger effect than tunneling of the magnetic moment in ferromagnets. The Néel vector, however, does not couple to any experimental

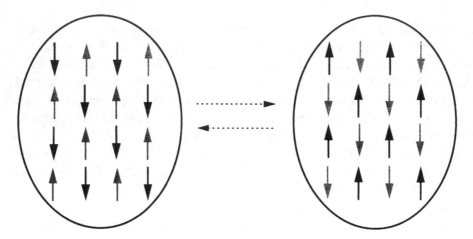

Figure 3.11   Quantum tunneling of antiferromagnetic sublattices.

field. Consequently, the antiferromagnetic tunneling can be detected by magnetic measurements only if there is a small magnetic moment due to the non-compensation of sublattices. In antiferromagnetic grains the biggest non-compensation arises from the different sizes of sublattices due to the irregular shape of the grain. There can also be a small non-compensation due to the canting of sublattices. This is known as weak ferromagnetism.

Consider a small antiferromagnetic grain having two collinear ferromagnetic sublattices whose magnetizations, m_1 and m_2, are coupled by the strong exchange interaction $\chi_\perp^{-1} m_1 \cdot m_2$ $(\chi_\perp \ll 1)$. We shall assume a small non-compensation of sublattices: $m_1 > m_2$, $m = m_1 - m_2 \ll m_1$. For a single-domain grain of volume V, only total moments of sublattices, $M_{1,2} = m_{1,2} V$, enter the problem. It is convenient, however, to keep the explicit volume dependence of all variables. Without anisotropy terms the magnetic Lagrangian of the grain is

$$\mathcal{L}_0 = V\left(\frac{m_1}{\gamma}\dot{\phi}_1(\cos\theta_1 - 1) + \frac{m_2}{\gamma}\dot{\phi}_2(\cos\theta_2 - 1)\right.$$
$$\left. - \chi_\perp^{-1} m_1 m_2 [\sin\theta_1 \sin\theta_2 \cos(\phi_1 - \phi_2) + \cos\theta_1 \cos\theta_2 + 1]\right), \quad (3.54)$$

where γ is the gyromagnetic ratio, and $\theta_{1,2}$ and $\phi_{1,2}$ are the spherical coordinates of constant-length vectors m_1 and m_2. To obtain the rate of quantum interchange of m_1 and m_2, one should compute the path integral

$$\int \mathcal{D}\{\theta_1\}\,\mathcal{D}\{\theta_2\}\,\mathcal{D}\{\phi_1\}\,\mathcal{D}\{\phi_2\}\,\exp\left(\frac{i}{\hbar}\int dt\,\mathcal{L}\right) \qquad (3.55)$$

over all trajectories that lead from the initial to the final state. Here $\mathcal{L} = \mathcal{L}_0 + \mathcal{L}_1$ is the total Lagrangian, which includes relativistic corrections to Eq. (3.54) due to the magnetic anisotropy.

In an experiment one looks for tunneling of the excess moment, $mV = (m_1 + m_2)V$. We should, therefore, single out the coordinates of m, and integrate first over the angular variables which are not directly measured. Since we are looking for quantum transitions between macroscopic states, only low-energy trajectories with almost antiparallel m_1 and m_2 contribute to the path integral. It is, therefore, safe to say that tunneling of m follows tunneling of m_1. For that reason we can replace θ_2 and ϕ_2 by $\pi - \theta_1 - \epsilon_\theta$ and $\pi + \phi_1 + \epsilon_\phi$, respectively, (with $|\epsilon_\theta|, |\epsilon_\phi| \ll 1$) in $\mathcal{L}_0$, and put $\theta_2 = \pi - \theta_1$ and $\phi_2 = \pi + \phi_1$ in $\mathcal{L}_1$. To second order in ϵ, Eq. (3.54) becomes

$$\mathcal{L}_0 = V\left(-\frac{m_1 + m_2}{\gamma}\dot{\phi}_1 + \frac{m}{\gamma}\dot{\phi}_1\cos\theta_1 + \frac{m_2}{\gamma}\epsilon_\theta\dot{\phi}_1\sin\theta_1 - \frac{m_2}{\gamma}(\epsilon_\phi\sin\theta_1)\dot{\theta}_1\right.$$
$$\left. - (2\chi_\perp)^{-1}m_1m_2\epsilon_\theta^2 - (2\chi_\perp)^{-1}m_1m_2(\epsilon_\phi\sin\theta_1)^2\right). \tag{3.56}$$

Then the Gaussian integration over ϵ_θ and $\epsilon_\phi\sin\theta_1$ reduces Eq. (3.55) to

$$\int \mathcal{D}\{\theta\}\,\mathcal{D}\{\phi\}\exp\left(\frac{i}{\hbar}\int dt\,\mathcal{L}_{\text{eff}}\right), \tag{3.57}$$

where we have replaced θ_1 and ϕ_1 by θ and ϕ, and introduced

$$\mathcal{L}_{\text{eff}} = V\left(-\frac{m_1 + m_2}{\gamma}\dot{\phi} + \frac{m}{\gamma}\dot{\phi}\cos\theta + \frac{\chi_\perp}{2\gamma^2}(\dot{\theta}^2 + \dot{\phi}^2\sin^2\theta)\right) + \mathcal{L}_1(\theta, \phi). \tag{3.58}$$

On the way from Eq. (3.54) to Eq. (3.58) we also put $m_1 = m_2$ in all terms except the first two, in which the difference between m_1 and m_2 is important.

To make the problem solvable, consider a tetragonal anisotropy,

$$\mathcal{L}_1 = -V[K_\perp\cos^2\theta + K_\parallel\sin^2\theta\sin^2\phi], \tag{3.59}$$

with $K_\perp \gg K_\parallel$. It corresponds to the X easy axis in the X–Y easy plane, Fig. 3.12. Let the initial state of the grain be m_1 in the positive X direction and m_2 in the negative X direction, that is, $\theta = \pi/2$ and $\phi = 0$ (Fig. 3.12). We are interested in the tunneling through the anisotropy barrier, $K_\parallel V$, into the state in which m_1 and m_2 have interchanged their orientations, that is, into the state in which $\theta = \pi/2$ and $\phi = \pi$. Taking into account that $K_\perp \gg K_\parallel$, the Gaussian integration over $\cos\theta$ in Eq. (3.57) reduces the path integral to

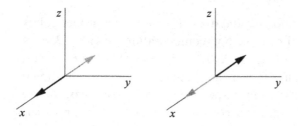

Figure 3.12 The geometry of the problem. Tunneling occurs between the two states shown. Black and white arrows indicate the moments of the two sublattices.

$$\int \mathcal{D}\{\phi\} \exp\left[-\frac{V}{\hbar}\int d\tau \left(i\frac{m_1+m_2}{\gamma}\dot{\phi} + \frac{1}{2}(I_f+I_a)\dot{\phi}^2 + K_\parallel \sin^2\phi\right)\right],$$

(3.60)

where we have switched to the imaginary time $\tau = it$ (overdots now represent derivatives with respect to τ) and introduced the effective ferromagnetic and anti-ferromagnetic moments of inertia,

$$I_f = \frac{m^2}{2\gamma^2 K_\perp}, \quad I_a = \frac{\chi_\perp}{\gamma^2}.$$

(3.61)

The path integral is dominated by $\phi(\tau)$ close to the trajectories (instantons) which minimize the integral in the exponent of Eq. (3.60),

$$\phi = \pm 2\arctan\left[\exp\left(\omega_0\tau\right)\right],$$

(3.62)

where

$$\omega_0 = 2\gamma\left(\frac{K_\parallel K_\perp}{m^2 + 2\chi_\perp K_\perp}\right)^{1/2}.$$

(3.63)

These solutions of imaginary-time equations of motion carry out the clockwise and counterclockwise underbarrier rotation of sublattices from $\phi = 0$ at $\tau = -\infty$ to $\phi = \pm\pi$ at $\tau = +\infty$ (see Fig. 3.12). Before computing their contribution to the tunneling rate, let us discuss the first term in the exponent of Eq. (3.60). It generates the phase factor

$$\exp\left[-i(S_1 + S_2)\Delta\phi\right],$$

(3.64)

where $\Delta\phi$ is the total change in ϕ along the instanton. Here we have introduced the total spins of sublattices, $S_{1,2} = m_{1,2}V/(\hbar\gamma)$. The phases generated by the clockwise ($\Delta\phi = -\pi$) and counterclockwise ($\Delta\phi = \pi$) instantons combine into

$$\cos\left[(S_1 + S_2)\pi\right] = \cos\left(s\pi + 2S_2\pi\right),$$

(3.65)

which is 0 or ± 1 depending on whether s is an integer or a half-integer. This implies that, in the absence of the magnetic field, an integer s can tunnel but a half-integer s cannot [33,34]. Then, the conventional instanton method gives the following answer for the tunneling rate [39]:

$$\Gamma \approx |\cos(s\pi)|\omega_0 \exp\left[-\frac{2V}{\hbar\gamma}\left(2\chi_\perp K_\parallel + m^2\frac{K_\parallel}{K_\perp}\right)^{1/2}\right].$$

(3.66)

The effect of the non-compensation captured by this formula is rather robust with respect to the model [13]. For large non-compensation ($m \gg (\chi_\perp K_\perp)^{1/2}$) and for small non-compensation ($m \ll (\chi_\perp K_\perp)^{1/2}$), Eqs. (3.63) and (3.66) reduce to the expressions for tunneling in a ferromagnetic grain [4] and for tunneling in a compensated antiferromagnetic grain, respectively [6,40]. Note that, for any $m \neq 0$, no matter how small, $K_\perp \neq 0$ is needed in order to produce a non-zero tunneling rate. This is because, at $K_\perp = 0$, the excess spin is conserved exactly owing to its commutation with the Hamiltonian, and, therefore, cannot tunnel.

Tunneling can be observed if it dominates over thermal transitions. The temperature, T_c, of the crossover from quantum to thermal superparamagnetism may be roughly estimated by comparing the quantum exponent of Eq. (3.66) with the Boltzmann exponent $K_\parallel V/(k_B T)$. This gives $T_c \simeq \hbar\omega_0$. In almost compensated antiferromagnetic particles,

$$T_c \simeq \mu_B (H_\parallel H_{ex})^{1/2}, \qquad (3.67)$$

where we have introduced the anisotropy field $H_\parallel$ and the exchange field H_{ex}. Owing to the presence of the exchange in this formula, the crossover in antiferromagnets occurs at higher temperatures than it does in ferromagnets, even for modest values of the magnetic anisotropy. For $K_\parallel \simeq 10^6$ erg cm^{-3} and $\chi_\perp \simeq 10^{-4}$, T_c must be of the order of a few kelvins.

Another important observation is that, according to Eq. (3.66), for a compensated or nearly compensated antiferromagnetic particle, tunneling is much stronger than it would be in a ferromagnetic particle of the equivalent size. The crossover from strong antiferromagnetic to weak ferromagnetic quantum dynamics occurs at $m \simeq (\chi_\perp K_\perp)^{1/2}$. This means that, to insure antiferromagnetic dynamics, the relative non-compensation, m/m_1, must not exceed $(H_\perp/H_{ex})^{1/2}$. For typical values of the parameters this corresponds to a non-compensation of about 1%. In this case the tunneling exponent is of the order of $(V/\mu_B)(\chi_\perp K_\parallel)^{1/2} \simeq N s_0 (H_\parallel/H_{ex})^{1/2}$, where N is the number of magnetic atoms in the grain and s_0 is the atomic spin. The pre-exponential factor has the order of magnitude of $\omega_0 \simeq k_B T_c/\hbar$, that is, 10^{10}–10^{11} s^{-1}. Consequently, on the timescale of a typical experiment, tunneling must occur in antiferromagnetic particles containing 10^3–10^4 magnetic atoms without the fine tuning of the magnetic field needed in ferromagnetic particles. Ferritin particles (see Chapters 5 and 6), in which tunneling has been studied by resonance techniques and by magnetic relaxation, fall within this range because their hematite core contains a few thousand iron atoms.

3.5 Tunneling with dissipation, macroscopic quantum coherence, and decoherence

In recent years there have been published a few theoretical works that considered magnetic tunneling with dissipation due to phonons [41,42,29], conducting electrons [43], electromagnetic radiation [42], nuclear spins [10,11], etc. along the lines of the Caldeira–Leggett approach, see Chapter 2. Some of these effects are negligible but others deserve attention. The detailed account of the effect of interactions on magnetic tunneling is a topic as big as that of micromagnetism itself. It requires the extension of a substantial part of the conventional micromagnetic theory into imaginary time, which cannot be done in one book. For that reason we will limit our discussion to some general statements relevant to experiments.

For a small ferromagnetic or antiferromagnetic particle, the relative effect of the dissipation on the *real-time dynamics* of M is manifested by the width, $\Delta\omega$, of the ferromagnetic or antiferromagnetic resonance. The frequency of the ferromagnetic and antiferomagnetic resonances is always of the order of the corresponding instanton frequency ω_0 for the tunneling problem. Although the macroscopic equation of motion with dissipation for the magnetic moment, Eqs. (1.1) and (1.4), is not of the form studied by Caldeira and Leggett, for certain geometries and in the limit of weak dissipation it can be reduced to that form. According to Chapter 2, this leads one to believe that the relative contribution of the dissipation to the tunneling exponent is of order $\Delta\omega/\omega_0$. The relative width of the magnetic resonance can be as large as 0.1 in metals and as small as 10^{-6} in insulators. The subtle point, however, is that the instanton frequency goes to zero when the height of the barrier goes to zero. In the context of macroscopic quantum tunneling, we are interested in this very limit of a small barrier. Thus, even for systems with very narrow magnetic resonance, it is not clear beforehand that the effect of dissipation on tunneling will be small and one has to develop a more detailed approach based upon a particular mechanism of dissipation. We shall illustrate this point by an important example of the interaction of M with nuclear spins, given by Anupam Garg [10]. We will reformulate his ideas in terms of the parameters commonly used in the studies of nuclear magnetic resonance (NMR) [44].

The electronic, M, and nuclear, m, magnetizations are coupled via the hyperfine interaction

$$E_{hf} = AM \cdot m = -H_{hf} \cdot m = -h_{hf} \cdot M, \qquad (3.68)$$

where A is a constant. Here we have introduced the hyperfine field at the nucleus, $H_{hf} = -AM$, and the hyperfine field, $h_{hf} = -Am$, exerted by the nuclear moments on the electronic magnetization. The values of these fields for various materials are known from NMR studies. To a first approximation, the electronic sublattice satisfies

$$\dot{M} = \gamma M \times H_{eff}, \qquad (3.69)$$

where $H_{eff} = -\delta E/\delta M$ (E being the total magnetic energy).

For the purpose of illustration consider our model III in which the barrier is lowered by tuning the external magnetic field to the anisotropy field, $H_a = 2K_{\parallel}/M_0$. Taking account of Eq. (3.68), the effective field in that model becomes

$$H_{eff} = \epsilon H_a + h_{hf}, \qquad (3.70)$$

where $\epsilon = 1 - H/H_a$. The height of the barrier in that model is proportional to ϵ^2. From Eqs. (3.69) and (3.70) it is clear that the relative perturbation of the real- and imaginary-time dynamics of M by nuclear spins is of order $h_{hf}/(H_a\epsilon)$. Thus, the corresponding perturbation of the tunneling exponent is

$$B = B_0\left(1 + \frac{h_{hf}}{H_a\epsilon}\right). \tag{3.71}$$

The nuclear magnetization, to which h_{hf} is proportional, satisfies the Curie–Weiss law, so that the largest effect occurs at $T \ll \mu_n H_{hf}$ (μ_n being the moment of the nucleus). Table 3.1 summarizes relevant isotopes, their natural abundances, nuclear moments, nuclear ordering temperatures, and the zero-temperature values of the nuclear magnetization m, of the field h_{hf}, and of the ratio h_{hf}/H_a for five commonly used ferromagnetic elements [44]. The anisotropy field responsible for the barrier has been taken in the kilo-oersted range for Fe and Ni, and in the range of tens of kilo-oersteds for Co, Tb, and Dy, which should be a reasonable estimate for particles of size less than 100 Å. It follows from Table 3.1 that, for Fe- and Ni-based ferromagnetic materials, the interaction with nuclear spins becomes important for ϵ of order 10^{-4} and temperatures in the millikelvin range. In Co, the effect of nuclear spins must become noticeable already at $\epsilon \simeq 10^{-2}$ but only if the temperature is in the millikelvin range. The most pronounced effects should occur in Tb-based compounds because this element has a large nuclear moment and 100% concentration of the corresponding isotope.

The above consideration applies to the quantum decay of a metastable magnetic state. There is another interesting problem of magnetic tunneling that has been pursued actively in recent years. Let us, for the moment, forget about dissipation and imagine that, in a certain experiment, a doubly degenerate magnetic state is prepared that corresponds, e.g., to our model I. The tunneling removes the degeneracy by splitting the degenerate level into the true ground state and the first excited state separated by the energy

$$\Delta = \hbar\Gamma \tag{3.72}$$

from the ground state (Fig. 3.13). For macroscopic quantum tunneling the splitting Δ is always small compared with the energy of zero-point oscillations of M around one of its equilibrium orientations along the anisotropy axis. This is because the latter is of the order of the instanton energy, $\hbar\omega_0$, whereas $\Gamma \simeq \omega_0 \exp(-B)$ with $B \gg 1$. In that limit, if the spin Hamiltonian that provides the degeneracy is exact and no interactions of M with other degrees of freedom are involved, the wave functions of the true ground state and the first excited state can be approximated by linear combinations of the degenerate states in the absence of tunneling;

Table 3.1   Nuclear magnetism in fine particles of five commonly used ferromagnetic materials

Isotope	Abundance (%)	μ_n (nm)	$\mu_n H_{hf}$ (mK)	m (emu cm^{-3})	h_{hf} (Oe)	h_{hf}/H_a
^{57}Fe	2.2	0.090	1.1	8.11×10^{-4}	0.16	$\simeq 10^{-4}$
^{59}Co	100	4.6	38	2.1	330	$\simeq 10^{-2}$
^{61}Ni	1.2	0.75	2.0	4.0×10^{-3}	0.59	$\simeq 10^{-4}$
^{159}Tb	100	2.0	220	0.31	360	$\simeq 10^{-2}$
^{161}Dy	19	0.46	99	1.4×10^{-2}	27	$\simeq 10^{-3}$
^{163}Dy	25	0.64	130	2.6×10^{-2}	49	$\simeq 10^{-3}$

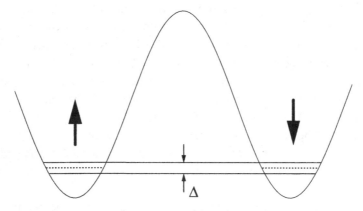

Figure 3.13   Tunneling splitting of a doubly degenerate spin state.

$$|0\rangle = \frac{1}{\sqrt{2}}(|\uparrow\rangle + |\downarrow\rangle)$$

$$|1\rangle = \frac{1}{\sqrt{2}}(|\uparrow\rangle - |\downarrow\rangle). \tag{3.73}$$

If a single-domain particle is in the true ground state, the probability of its magnetic moment having a certain orientation, M, at a moment of time t oscillates with time,

$$\langle M(t)\cdot M(0)\rangle = M_0^2 \cos(\Gamma t). \tag{3.74}$$

Such a system placed in an AC field of frequency ω would show a resonance in the power absorption at $\omega = \Delta/\hbar = \Gamma$. This effect has become known as macroscopic quantum coherence [45]. Its observation in magnetic particles would be a spectacular manifestation of quantum mechanics at a macroscopic or, at least, mesoscopic level. There are certain difficulties, however, in the way of such an experiment.

The most serious obstacle for the observation of macroscopic quantum coherence is that the interaction of M with the environment of order Δ will destroy the coherent oscillations and the resonance in the absorption spectrum. This is radically different from the condition of the small effect of dissipation on the tunneling rate. The latter condition is satisfied if the characteristic energy of the interaction of M with the environment is small compared with the instanton energy, $\hbar\omega_0$, which is much greater than Δ. Insofar as the condition for coherence is concerned, it is clear that the more macroscopic the coherence (that is, the greater B) the more fragile it is with respect to the interactions. If, e.g., the tunneling rate were as small as $\Gamma \simeq 1 \text{ s}^{-1}$, which one can easily imagine for a macroscopic system, the interaction required to destroy the coherence would be as negligible as 10^{-15} eV. In that case, interactions of M with fluctuating external fields, phonons, conducting electrons, nuclear spins, etc., must destroy the coherence. We have seen, however, that Γ, in almost compensated antiferromagnetic particles, can be quite large even for particles of a considerable size. Thus, antiferromagnetic particles would be the first candidates for measurements of macroscopic quantum coherence. Let us discuss the feasibility of such an experiment.

First of all the system must be shielded from external fields of all possible origins. At present this can be done down to 10^{-6} Oe. There are, however, intrinsic fields about which we have to be concerned. We have seen from the discussion of nuclear spins that a non-zero nuclear moment exerts the hyperfine field h_{hf} on the electronic magnetization. At low temperature, when nuclear spins are ordered, this field for the materials listed in Table 3.1 is in the range 0.2 Oe to 0.4 kOe. This is more than enough to completely destroy the coherence. One can argue that for iron, e.g., the nuclear spins are disordered above 1 mK, so that the corresponding h_{hf} must be lower than the listed value of about 0.2 Oe. How much lower is it? Without any isotopic purification, a small iron particle consisting of 10^4 atoms must have about $N_{\text{n}} \simeq 200$ ^{57}Fe nuclei. If their spins are thermally disordered there still should be some net nuclear moment, M, due to statistical fluctuations. On average, it will produce $h_{\text{hf}} \simeq 0.2/\sqrt{N_n} \simeq 10^{-2}$ Oe. This field will destroy the coherence unless $\Gamma > \gamma h_{\text{hf}} \simeq 10^5$. Anupam Garg [10] made the interesting observation that coherence will nevertheless be possible for short periods when the nuclear subsystem is in a microstate with $m = 0$. The probability for this to happen is

$$P_0 = (2\pi N_{\text{n}})^{-1/2}\left[\cosh\left(\frac{\mu_{\text{n}} H_{\text{hf}}}{T}\right)\right]^{-N_{\text{n}}}. \tag{3.75}$$

This gives the factor by which the power absorption in a resonance experiment must decrease owing to nuclear spins.

Despite having pointed out the above difficulties, we believe that the coherent quantum oscillations of the magnetic moment, similarly to the quantum oscillations of the ammonia molecule [46], can be observed. The question is that of how macroscopic they are going to be. At present, it appears that the observation of macroscopic quantum coherence in a system of 100 or fewer, antiferromagnetically coupled, spins is a feasible experiment whereas in a bigger system it may require great experimental effort: isotopic purification, shielding down to very weak fields, very low temperatures, etc. If the conditions for coherence are not satisfied, the system will tunnel incoherently in a random manner between the wells shown in Fig. 3.13. Note that tunneling, in contrast to coherence, is much more robust with respect to the dissipation.

Chapter 4

Magnetic tunneling in bulk materials

Until now we have studied problems of homogeneous spin tunneling in small particles, in which we assumed the magnetic moment (or moments of sublattices) to be independent of spatial coordinates. In that case, as we have seen, the tunneling exponent is always proportional to the total tunneling spin. The particle, therefore, must be sufficiently small to provide a significant tunneling probability. When we switch to considering tunneling in bulk magnets, the same kind of argument can be made that suggests that magnetic tunneling can only occur in a small volume comparable to that in single-domain particles. A significant difference, however, is that a local tunneling event in a bulk magnet can trigger instability on a much greater scale, leading to really macroscopic consequences. Examples are quantum nucleation of magnetic domains and quantum depinning of domain walls, which are studied below. At the end of this chapter we will also briefly discuss tunneling of magnetic flux lines in superconductors.

4.1   Quantum nucleation of magnetic domains

Nucleation of magnetic bubbles in thin ferromagnetic films is an interesting fundamental problem that may have great potential for experimental studies. Consider a thin film of ferromagnetic material with strong perpendicular anisotropy, which is uniformly magnetized in the positive Z-direction, Fig. 4.1. The magnetic field is applied in the negative Z-direction, which favors the reversal of the magnetization. The reversal occurs via the nucleation of a critical bubble that then grows in size until the magnetization of the whole film has become aligned with the direction of the field. In many bubble systems the magnetic

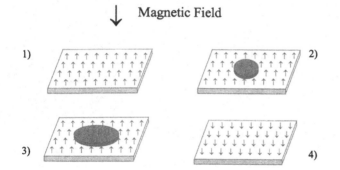

dipole energy of the film is crucial for the understanding of this process. Here we will simplify the problem by considering systems in which the energy of the magnetic anisotropy is large compared with the dipole energy. In this case the dipole energy can be neglected and the nucleation is entirely determined by the magnetic anisotropy and the exchange interaction. At high temperature, the nucleation of a bubble is a thermal overbarrier process, which goes with the rate $\Gamma \propto \exp(-U/T)$, where U is the barrier for the formation of the critical nucleus. As the temperature decreases, thermal transitions die out and the nucleation may occur only due to quantum tunneling [5]. The initial uniform state of the film and the state immediately after the bubble has nucleated are macroscopically distinct. This puts the problem within the domain of macroscopic quantum tunneling. Our purpose is to generalize the results of the previous chapter to the case of the non-uniform rotation of the magnetic moment. Quantum nucleation, being a field theory problem, is more difficult than tunneling of M in single-domain particles. In an experiment, however, it may be easier to monitor single nucleation events in a thin film than to detect the magnetization reversal in a nanometer-size particle.

The rate of the decay of a metastable state is given by Eq. (2.12), where the partition function is now the integral over the magnetization field periodic in τ, $M(r, \tau)$,

$$Z = \oint \mathcal{D}\{M(r, \tau)\} \exp\left(-\frac{1}{\hbar}\int d\tau \int d^3r\, \mathcal{L}_E\right). \tag{4.1}$$

Here $\mathcal{L}_E = -L(it = \tau)$ is the Euclidean version of the density of the magnetic Lagrangean,

$$\mathcal{L} = \left(\frac{M_0}{\gamma}\right)\dot{\phi}(\cos\theta - 1) - E(\theta, \phi), \tag{4.2}$$

where E is now the energy density and both angles depend on coordinates and time. The integration over $\tau = it$ in Eq. (4.1) is performed over $M(r, \tau)$ trajectories that are periodic in τ with the period $\hbar/T$. As previously, we are making the traditional assumption that the local spin density is determined by the strong ferromagnetic exchange and cannot be changed. This translates into the constant length, M_0, of the vector M, which makes the description in terms of the magnetization field equivalent to the description in terms of the two fields $\theta(r, \tau)$ and $\phi(r, \tau)$.

The energy density of the ferromagnet consists of three parts,

$$E = E_a(\theta, \phi) + E_e(\theta, \phi) + E_m,\tag{4.3}$$

where E_a, E_e, and E_m are the magnetic anisotropy energy, the exchange energy, and the Zeeman energy, respectively. We shall assume for simplicity that all these energies are large compared with the magnetic dipole energy, which is, therefore, neglected. The energy density due to the external field is

$$E_m = -M \cdot H = M_0 H \cos \theta,\tag{4.4}$$

where we have taken into account that H is applied in the negative Z-direction (see Fig. 4.1). The exchange energy density is

$$E_e = \frac{\alpha}{2} (\partial_i M_j)^2 = \frac{\alpha}{2} M_0^2 [(\nabla \theta)^2 + \sin^2 \theta \, (\nabla \phi)^2],\tag{4.5}$$

where α is the exchange stiffness constant, which is typically of order 10^{12} cm^2. This term in the total energy reflects the fact that (in the absence of the dipole energy) the minimum of the energy corresponds to uniform magnetization of the ferromagnet. Finally, we will select the anisotropy energy density in the minimal form that allows quantum nucleation:

$$E_a = \text{const} - k_\parallel M_z^2 + k_\perp M_y^2 = (K_\parallel + K_\perp \sin^2 \phi) \sin^2 \theta,\tag{4.6}$$

where $K_\parallel$ and $K_\perp$ are anisotropy constants in erg cm^{-3}. This is actually the anisotropy of our model III, where the axes X and Z have been interchanged. It describes the film with the X–Z easy plane and the Z easy direction in that plane. The two uniformly magnetized states of minimal energy are $\theta = 0$ and $\theta = \pi$. The non-zero $K_\perp$ is responsible for quantum transitions between these two states. If $K_\perp = 0$, M_z commutes with E_a and is conserved exactly. The saddle point at $\phi = 0$ corresponds to $\cos \theta_1 = H/H_c$, where $H_c = 2K_\parallel/M_0$. The energy barrier for the nucleation exists at $H < H_c = 2K_\parallel/M_0$. A small barrier can be achieved by tuning H to H_c. Since we are interested precisely in that case of a small barrier, it is convenient to introduce a small parameter $\epsilon = 1 - H/H_c$. This simplifies the mathematics tremendously because only small θ become involved in the nucleation process.

At $\epsilon \ll 1$ the functional integration over ϕ becomes Gaussian and can be performed exactly in a manner similar to the integration over θ in model III of Chapter 3. The remaining integral is of the form

$$\oint \mathcal{D}\{\theta(r, \tau)\} \exp\left(-\frac{I_{\text{eff}}}{\hbar}\right),\tag{4.7}$$

where the temperature-dependent effective Euclidean action is given by

$$I_{\text{eff}} = C \int d\bar{\tau} \int d^2\bar{r} \left[\frac{1}{2} \left(\frac{\partial \bar{\theta}}{\partial \bar{\tau}} \right)^2 + \frac{1}{2} (\bar{\nabla}\bar{\theta})^2 - \frac{1}{2} (\bar{\theta}^2 - 1)^2 \right]. \qquad (4.8)$$

Here $C = 8K_{\parallel}\delta^2 d(2\epsilon)^{1/2}/\omega$ is a field-dependent constant, d is the thickness of the film, and the following dimensionless variables have been introduced:

$$\bar{\tau} = \tau\omega(\epsilon/2)^{1/2}, \quad \bar{r} = r(\epsilon/2)^{1/2}/\delta, \quad \bar{\nabla} = \partial/\partial\bar{r}, \quad \bar{\theta} = \theta/(2\epsilon)^{1/2}. \qquad (4.9)$$

The parameters δ and ω are defined as $\delta = \left[\alpha M_0^2/(2K_{\parallel}) \right]^{1/2}$ and $\omega = (2\gamma/M_0)(K_{\parallel}K_{\perp})^{1/2}$. Note that δ is the thickness of the domain wall in a bulk ferromagnet.

We will give the solution of the nucleation problem for an arbitrary temperature. The idea underlying that calculation has been discussed in Chapter 2. The extrema of I_{eff} within the class of functions $\bar{\theta}(\bar{r}, \bar{\tau})$, which are periodic in $\bar{\tau}$ with the period $P = \hbar\omega\epsilon^{1/2}/T$, determine the main contribution to the path integral at temperature T. The problem then reduces to the computation of $T \neq 0$ thermons (according to the terminology of Chapter 2) of the non-linear scalar field model in $2 + 1$ dimensions. They are solutions of the equation of motion derived from the action (4.8)

$$\frac{\partial^2 \bar{\theta}}{\partial \bar{\tau}^2} + \frac{\partial^2 \bar{\theta}}{\partial \bar{r}^2} + \frac{1}{\bar{r}} \frac{\partial \bar{\theta}}{\partial \bar{r}} = 2(\bar{\theta} - \bar{\theta}^3), \qquad (4.10)$$

where $\bar{r} = |\bar{r}|$. The solutions must obey $\bar{\theta} \to 0$ as $\bar{r} \to \infty$ (in order to have finite action), and must be periodic in $\bar{\tau}$, with $\partial\bar{\theta}/\partial\bar{\tau} = 0$ at $\bar{\tau} = 0$ and $\bar{\tau} = \pm P/2$. The numerical solution of the above equations presents a certain mathematical challenge. The reader interested in the details can find them in [47]. Here we will only comment on the results.

The shape of the various thermons is illustrated in Fig. 4.2. At zero temperature (i.e., infinite period), the dominant solution is $O(2)$ symmetric in $\bar{\tau}$ and $\bar{r}$, $\bar{\theta} = \bar{\theta}(\bar{r}^2 + \bar{\tau}^2)$. This is the usual instanton of the $2 + 1$ field theory with the potential $\bar{\theta} - \bar{\theta}^3$ (Fig. 4.2a). It is important to notice the fast exponential decay of the $T = 0$ solution for values of $\bar{r}$ outside the peak at its center. At $T \neq 0$ the thermon must have a period $P = \hbar\omega\epsilon^{1/2}/T$ on $\bar{\tau}$. At low temperature, when this period is large, the solution is close to the periodic arrays of the $T = 0$ instantons in the $\bar{\tau}$ direction, with their centers weakly interacting owing to the exponentially small overlap. This, in fact, remains true until the temperature becomes quite close to $T_c = \epsilon^{1/2}\hbar\omega$. As the temperature approaches T_c, the thermon acquires a 'sierra' shape (Fig. 4.2b,c), with the 'valleys' gradually rising to the height of the 'peaks'. At $T = T_c$ oscillations on τ flatten out, producing the τ-independent 'ridge' along the τ-axis in the τ–r plane. This 'ridge' corresponds to the thermal nucleation of the bubble. Note that the bifurcation of the thermon of Eq. (4.10) at $T = T_c$ smoothly transforms the 'sierra' into the 'ridge' (Fig. 2.14). Consequently the transition from quantum to thermal nucleation is of second order (see Chapter 2), although it occurs within a rather narrow temperature range.

T = 0

(a)

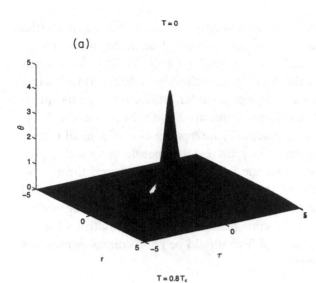

T = 0.8 T_c

(b)

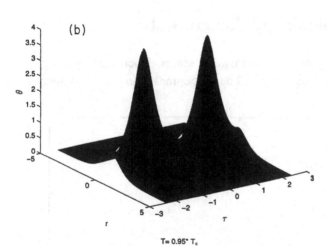

T = 0.95* T_c

(c)

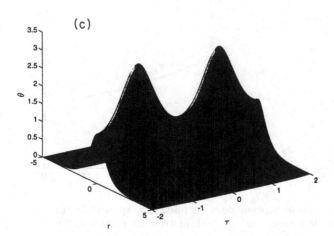

Figure 4.2 Thermons corresponding to the nucleation of the magnetization reversal bubble in a thin ferromagnetic film: (a) $T = 0$, (b) $T = 0.8T_\mathrm{c}$, and (c) $T = 0.95T_\mathrm{c}$.

Figure 4.3 presents the temperature dependence of the Euclidean action (the tunneling exponent) obtained by the numerical integration of the thermon solution of Eq. (4.10). The value of the $T = 0$ integral in Eq. (4.8) is 13.4. As expected from the smooth transition of the thermon from the 'sierra' to the 'ridge' shape, the graph in Fig. 4.3 shows a smooth ('second-order') transition from the quantum to the thermal regime. It should be emphasized that the curve in Fig. 4.3 is universal in the sense that, in the practically interesting case of a small barrier (small ϵ), it is exclusively determined by the $2 + 1$ dimensionality of the problem. This opens an interesting avenue for comparison between theory and experiment. In such an experiment the height of the barrier should be controled by fine tuning of the magnetic field within a certain area of the film. The nucleation should occur each time at the same place to provide the statistics for the nucleation rate. The temperature and field should be good control parameters for testing predictions of the theory.

4.2   Quantum depinning of domain walls

Tunneling exists in the presence of metastable states. There are two major sources of metastability in ferromagnets. The first comes from the fundamental

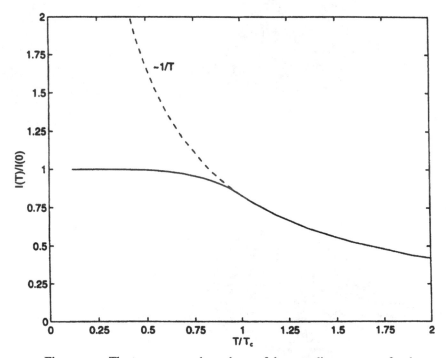

Figure 4.3   The temperature dependence of the tunneling exponent for the nucleation of the magnetization reversal bubble in a thin ferromagnetic film.

non-linearity of the magnetic energy. It leads to the existence of several directions of easy magnetization and to the appearence of magnetic domains. Up to now we have concentrated on tunneling out of such metastable states. The second source of metastability is various kinds of crystal defects. Owing to such defects the energy of a domain wall fluctuates over the volume of the sample. This leads to the pinning of walls by defects. In the presence of the magnetic field, domains whose magnetization is aligned with the field tend to grow, whereas domains whose magnetization is opposite to the field tend to collapse. The force on the domain wall is proportional to the field. As the field grows, this force eventually exceeds the pinning force and the wall becomes mobile. However, even well below the coercive field, one usually observes some slow dynamics of the magnetization owing to the thermal overbarrier diffusion of domain walls in the pinning potential. As usual, thermal transitions die out as the temperature goes to zero. Here we are interested in quantum depinning of domain walls. Two conditions must be satisfied to make quantum effects observable. First, the rate of quantum tunneling must be significant, which requires a small barrier. Secondly, the temperature of the crossover from the thermal to the quantum regime must be accessible experimentally. Experiments on quantum depinning of domain walls may be very simple. The quantity to be measured is the rate of the relaxation of the magnetization at low temperature. A temperature-independent relaxation rate below some temperature would be strong evidence for the quantum nature of the relaxation process. In fact, quantum diffusion of microscopically thin domain walls, separating two atomic planes with opposite spin orientations, was proposed quite some time ago [48] to explain temperature-independent magnetic after-effects [49,50] observed in highly anisotropic magnets at low temperature. Such walls exist only in materials with the highest known magnetic anisotropy values. Their dynamics are microscopic rather than macroscopic. Typical walls have a thickness of a few hundred Å.

A domain wall corresponds to the rotation of the magnetization vector from one magnetic domain to another. Tunneling of the wall is, thus, a transition between two $M(x)$ configurations. Finding the instanton of Eq. (1.1) for a wall tunneling through the pinning barrier is rather difficult. Instead, we will treat the domain wall as a two-dimensional object characterized by the metric g_{ik} or by the equation of the surface $Z(x, y, t)$. In this approach the dependences of the width and the energy of the wall on its velocity and orientation will be left out of the picture. We will demonstrate, however, that our approximation is rather good in the limit of a small barrier, which is of primary interest for the tunneling problem. Our main aim will be the calculation of the tunneling rate and the crossover temperature in terms of macroscopic parameters of the system.

We will treat the domain wall as a surface (Fig. 4.4) of energy Σ per unit area. In general the energy density $\Sigma(P,R)$ is a functional of the generalized coordinates $P(\xi)$ (the momentum density) and $R(\xi)$ (the radius vector of the wall);

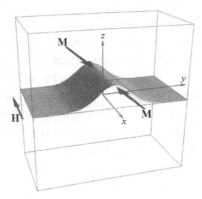

Figure 4.4 Decoupling of the domain wall from a planar defect located in the $X-Y$ plane.

here $\xi = (\xi_0, \xi_1, \xi_2)$ parametrizes the wall surface, with ξ_0 representing time, and (ξ_1, ξ_2), the spatial coordinates on the wall. In general, the dynamics of the wall is governed by the Slonczewski equations [51], which take the form

$$\frac{\delta \Sigma}{\delta \boldsymbol{R}(\xi)} = -\dot{\boldsymbol{P}}(\xi)\,, \quad \frac{\delta \Sigma}{\delta \boldsymbol{P}(\xi)} = \dot{\boldsymbol{R}}(\xi). \tag{4.11}$$

For an arbitrary wall these can be very complicated, but for the gently curved wall of Fig. 4.4 one has a wall energy density $\Sigma(\boldsymbol{P}, \boldsymbol{R})$ of the form

$$\Sigma = \sigma \left[\left(1 + \frac{\gamma^2 \delta^2}{4 M_0^2} (\nabla \boldsymbol{P}_\perp)^2 + \frac{\dot{Z}^2}{v_0^2 [1 + (\nabla Z)^2]} \right) \left(1 + (\nabla Z)^2 \right) \right]^{1/2} \tag{4.12}$$

where $\boldsymbol{P}_\perp(x, y, t)$ is the component of the wall momentum density perpendicular to the wall surface $Z(x, y, t)$, δ is the wall thickness, v_0 is the limiting (Walker) velocity of the wall, and σ is the constant energy density of a planar stationary wall.

The term proportional to $(\nabla \boldsymbol{P}_\perp)^2$ in Eq. (4.12) corresponds to an inhomogeneous exchange energy in the wall. It has the order of δ^2 / R_0^2 compared with the term $\dot{Z}^2 / v_0^2$, R_0 being the curvature radius of the wall. The main approximation we shall make is to assume weak curvature ($R_0 \gg \delta$) and slow ($|\dot{Z}| \ll v_0$) motion of the wall. The equations of motion (4.12) then become equivalent to the covariant action,

$$I_0 = -\sigma \int \mathrm{d}^3 \xi \sqrt{g(\xi)}, \tag{4.13}$$

where $g(\xi) = \det [g_{ij}(\xi)]$ and g_{ij} is the metric tensor,

$$g_{ij} = \frac{\partial \boldsymbol{R}(\xi)}{\partial \xi_i} \cdot \frac{\partial \boldsymbol{R}(\xi)}{\partial \xi_j}. \tag{4.14}$$

To Eq. (4.13) we should also add the terms arising from a pinning potential $U[R(\xi)]$ and a volume magnetic energy arising from the magnetic field H; this gives a further contribution $I_p + I_m$ to the action, where

$$I_p + I_m = -\int d^3\xi \sqrt{g(\xi)} U[R(\xi)] - \int dt \int d^3R\, M(R)\cdot H. \qquad (4.15)$$

It is convenient now to convert to the parametrization $\xi_0 = t$, $\xi_1 = x$, $\xi_2 = y$, and describe the wall by the single-valued function $Z(x, y, t)$ (note that, for cylindrical magnetic bubble nucleation, this would not be convenient; a better choice would be $Z = Z(\rho, \phi, t)$, with z, ρ, and ϕ being cylindrical coordinates). With the approximations described above, we can then rewrite Eq. (4.13) as

$$I_0 = -\sigma \int dt\, dx\, dy \left(1 + (\nabla Z)^2 - \frac{1}{v_0^2}\dot{Z}^2\right)^{1/2}, \qquad (4.16)$$

The physical meaning of Eq. (4.16) becomes obvious in the limit of a small velocity of the wall, $|\dot{Z}| \ll v_0$. By expanding the square root under the integral, one obtains

$$I_0 = \int dt \int \frac{dx\, dy}{\cos\theta} \frac{m_0 v_\perp^2}{2} - \int dt \int \frac{dx\, dy}{\cos\theta} \sigma, \qquad (4.17)$$

where $\theta(x, y, t)$ is the angle that the vector normal to the surface makes with the Z-axis, $v_\perp(x, y, t) = \dot{Z}\cos\theta$ is the local velocity of the wall, and $m_0 = \sigma/v_0^2$ is the mass of a unit area of the wall. Here $dx\, dy / \cos\theta$ is the element of the surface, $\cos\theta = [1 + (\partial_x Z)^2 + (\partial_y Z)^2]^{-1/2}$. Correspondingly, the two terms in Eq. (4.17) represent the kinetic and the surface energy contributions to the action of the wall.

Let us now switch to imaginary time and use dimensionless variables,

$$\begin{aligned} x_0 = \omega\tau, \quad x_1 = x/\delta, \quad x_2 = y/\delta \\ \bar{z} = Z/\delta, \quad u = U(x, y, z)/\sigma, \quad h = 2M_0 H\delta/\sigma, \end{aligned} \qquad (4.18)$$

where δ is the domain wall thickness, $\omega = v_0/\delta$. Then the total Euclidean action becomes

$$I_E = -\frac{\sigma\delta^2}{\omega} \int d^3x \left\{(1 + [\nabla\bar{z}(x)]^2)^{1/2} + u(x_1, x_2, \bar{z}) - h\bar{z}(x)\right\}, \qquad (4.19)$$

where $x = (x_0, x_1, x_2)$ and $\nabla = (\partial_0, \partial_1, \partial_2)$.

The parameters of micromagnetic theory, which represent the macroscopic properties of ferromagnets, are exchange, A_e (erg cm^{-1}), and anisotropy, K (erg cm^{-3}), constants. It may be helpful to have the dependence of our parameters σ, δ, and ω on these constants. We will illustrate it by the example of a rhombic crystal with the magnetic anisotropy energy characterized by two positive constants, $K_\parallel$ and $K_\perp$:

$$E_a = -K_\parallel \frac{M_x^2}{M_0^2} + K_\perp \frac{M_z^2}{M_0^2}. \tag{4.20}$$

This describes a magnet having the X-axis as the easy magnetization direction, and the X–Y plane as the easy plane. Consequently, the magnetization inside the domain wall rotates in the X–Y plane (see Fig. 4.4). For such a wall

$$\sigma = 4(A_e K_\parallel)^{1/2}, \quad \delta = (A_e/K_\parallel)^{1/2}. \tag{4.21}$$

The limiting velocity is given by (see, e.g., [51])

$$v_0 = \frac{2\gamma}{M_0} (A_e K_\parallel)^{1/2} \left[\left(1 + \frac{K_\perp}{K_\parallel} \right)^{1/2} - 1 \right]. \tag{4.22}$$

Note that, in the limit of a uniaxial crystal ($K_\perp = 0$), the wall has an infinite mass, $m_0 = \sigma/v_0^2$, and, therefore, cannot move or tunnel. This is due to the fact that the Hamiltonian, $-K_\parallel \hat{s}_x^2$, produced in this limit by Eq. (4.20), commutes with the spin operator $\hat{s}_x$, so that s_x on each side of the wall becomes a conserved quantum number. In fact, however, the magnetic dipole interaction, which is not represented in Eq. (4.20), violates the commutation of the Hamiltonian with $\hat{s}_x$ even for a uniaxial crystal, leading to the replacement of $K_\perp$ by $2\pi M_0^2$. The frequency ω depends on anisotropy constants, but not on the exchange,

$$\omega = \gamma H_a \left[\left(1 + \frac{K_\perp}{K_\parallel} \right)^{1/2} - 1 \right], \tag{4.23}$$

where we have introduced the anisotropy field $H_a = 2K_\parallel/M_0$. As we will see below, $\hbar\omega/k_B$ determines the scale of the crossover temperature T_c.

Although we have not specified the pinning potential U, some important conclusions can be drawn concerning the magnitude of the reduced potential $u = U/\sigma$ in Eq. (4.19). The surface energy of the wall depends on the strength of the local exchange and anisotropy, Eq. (4.21). Fluctuations in A_e and $K_\parallel$ due to defects, and resulting fluctuations in σ, are responsible for the pinning. The wall is attracted to regions where σ is less than it is in an ideal crystal. Correspondingly, the value of σ in an ideal crystal is the energy available for pinning. This suggests that the reduced pinning potential, $u = U/\sigma$, satisfies $u < 1$. A similar argument can be made about the reduced field h. From Eq. (4.21) we get $h = H/H_a$. Energy barriers in ferromagnets normally exist at $H < H_c \le H_a$, which gives $h \le 1$.

Equation (4.19) allows one to compute the WKB exponent for an arbitrary curvature of the wall and an arbitrary pinning potential. We will concentrate on the case of small curvature and 'slow' instantons, $(\nabla \bar{z})^2 \ll 1$, the significance of which will be explained below. To simplify the problem we will study *planar defects* only, for which $u = u(z)$. The latter may be justified by the fact that, for energetic reasons, walls tend to be matched with planar defects or with groups of defects concentrated in one plane. One can also imagine a specific geometry of the experiment such that the wall tunnels through a planar defect. In the limit of small space–time derivatives and $u = u(z)$, I_E reduces to [42]

$$I_E = -\frac{\sigma \delta^2}{\omega} \int d^3x \left(\tfrac{1}{2}(\nabla \bar{z})^2 + u(\bar{z}) - h\bar{z} \right). \tag{4.24}$$

4.2.1   The tunneling exponent in $1 + 1$ dimensions

Consider a flat domain wall of a small area A_w, tunneling through a planar defect of the same area. This may be treated as an approximation for the tunneling of a wall through a small defect, or may correspond to the situation in which the wall is coupled to a planar defect within a very thin wire, A_w being the cross-section of the wire. Then Eq. (4.24) becomes

$$I_E = -\frac{\sigma A_w}{\omega} \int dx_0 \left[\tfrac{1}{2}\left(\frac{d\bar{z}}{dx_0} \right)^2 + u(\bar{z}) - h\bar{z} \right]. \tag{4.25}$$

The extremal trajectory satisfies

$$\frac{d^2\bar{z}}{dx_0^2} = \frac{du}{d\bar{z}} - h. \tag{4.26}$$

The total potential, $u(\bar{z}) - h\bar{z}$, of the general form is shown in Fig. 4.5a. The wall tunnels from $\bar{z} - z_1$ to $\bar{z} = z_2$. As is shown in Fig. 4.5b, z_1 and z_0 are solutions of $du/d\bar{z} = h$. At $h = h_c$ the energy barrier disappears. This corresponds to $z_2 \to z_0 \to z_1 \to 0$ at $h \to h_c$. In the case of a small barrier ($h \to h_c$), the general form of $du/d\bar{z}$ is (see Fig. 4.5b)

$$\frac{du}{d\bar{z}} = h_c - \tfrac{1}{2}\left(\frac{\bar{z} - r}{\bar{w}} \right)^2, \tag{4.27}$$

where $\bar{w} = w/\delta$ and w is the parameter characterizing the width of the barrier. Eq. (4.27) gives

$$z_1 = r - \bar{w}(2h_c\epsilon)^{1/2}, \quad z_0 = r + \bar{w}(2h_c\epsilon)^{1/2}, \tag{4.28}$$

where we have introduced a small parameter $\epsilon = 1 - h/h_c$. It is convenient to choose the Z-axis such that $z_1 = 0$. Then $r = \bar{w}(2h_c\epsilon)^{1/2}$, $z_0 = 2\bar{w}(2h_c\epsilon)^{1/2}$, and the total potential takes the form

$$u(\bar{z}) - h\bar{z} = \frac{1}{2\bar{w}}(2h_c\epsilon)^{1/2}\bar{z}^2 - \frac{1}{6\bar{w}^2}\bar{z}^3. \tag{4.29}$$

Consequently, $z_2 = 3\bar{w}(2h_c\epsilon)^{1/2}$.

With the help of Eq. (4.26) the extremal action can be presented as

$$I_E = -\frac{\sigma A_w}{\omega} \int_{z_1}^{z_2} d\bar{z} \left\{ 2[u(\bar{z}) - h\bar{z}] \right\}^{1/2}. \tag{4.30}$$

On substituting here the potential of Eq. (4.29), we obtain

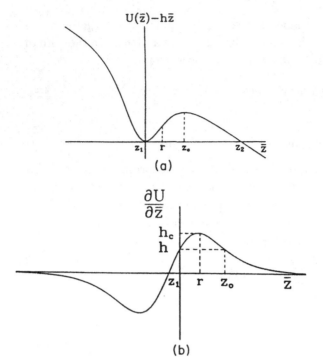

$U(\bar{z}) - h\bar{z}$

(a)

$\dfrac{\partial U}{\partial \bar{z}}$

(b)

Figure 4.5   (a) The shape of the potential, $u_0(z) - hz$. (b) $\partial u_0/\partial z$ versus z.

$$I_E = -\frac{48}{5}\frac{\sigma A_w}{\omega}\,\bar{w}^{3/2}(h_c\epsilon)^{5/4}. \tag{4.31}$$

The tunneling rate can be presented as

$$P = A\exp(-B_0), \tag{4.32}$$

where the WKB exponent is given by $B_0 = -I_E/\hbar$. In terms of the total number of tunneling spins, $N = M_0 A_w \delta/\mu_B$, with the help of Eqs. (4.18) and (4.31), we obtain [42]

$$B_0 = \frac{48}{5}\frac{\gamma H_c}{\omega}\left(\frac{w}{\delta}\right)^{3/2}h_c^{1/4}\epsilon^{5/4}N. \tag{4.33}$$

Let us now verify the validity of our assumption that $(d\bar{z}/dx_0)^2 \ll 1$, under which Eq. (4.24) was obtained from Eq. (4.19). The first integral of Eq. (4.26) is

$$\tfrac{1}{2}\left(\frac{d\bar{z}}{dx_0}\right)^2 = u(\bar{z}) - h\bar{z}. \tag{4.34}$$

According to Fig. 4.5a and Eq. (4.29) it has a maximum at $\bar{z} = z_0$,

$$\left(\frac{d\bar{z}}{dx_0}\right)^2_{\max} = \frac{8(h_c\epsilon)^{3/2}}{3\sqrt{2\bar{w}}}, \tag{4.35}$$

which is small in the limit of small ϵ and h_c.

Another interesting observation is the universal form of the instanton in the limit of a small barrier. By substituting Eq. (4.29) into Eq. (4.34), one obtains

$$\bar{z}(\tau) = z_2/\cosh^2(\omega_0\tau),\tag{4.36}$$

where

$$\omega_0 = \frac{2(2h_c\epsilon)^{1/4}}{\bar{w}^{1/2}}\,\omega.\tag{4.37}$$

Equation (4.36) describes the imaginary-time motion from $\bar{z} = 0$ at $\tau = -\infty$ to $\bar{z} = z_2$ at $\tau = 0$, and then back to $\bar{z} = 0$ at $\tau = +\infty$. It corresponds to the classical motion in the inverted potential, $h\bar{z} - u(\bar{z})$.

4.2.2 The tunneling exponent in $2+1$ dimensions

If the area of the domain wall coupled to a planar defect is large, the tunneling will occur via the nucleation process shown in Fig. 4.4. Once the critical portion of the wall has been released due to tunneling, the nucleus will expand and eventually depin the entire wall. Let X be the direction of the anisotropy axis and l be the length of the defect in that direction. The Euclidean action for that problem is

$$I_E = -\frac{\sigma l\delta}{\omega}\int d^2x\left(\tfrac{1}{2}(\nabla\bar{z})^2 + u(\bar{z}) - h\bar{z}\right),\tag{4.38}$$

where $d^2x = dx_0\,dx_2$ and $\bar{z} = \bar{z}(x_0, x_2)$. In this case it is impossible to solve the problem analytically, even for a small barrier. Instead, one can perform a dimensional analysis of Eq. (4.38), in order to extract the dependence of the WKB exponent on the parameters of the ferromagnetic material. Consider

$$z' = \frac{\bar{z}}{r},\quad x' = \frac{(2h_c\epsilon)^{1/4}}{\bar{w}1/2}\,x.\tag{4.39}$$

Then, with the help of Eq. (4.29), for a small barrier, we obtain

$$I_E = -\frac{\sigma l\delta}{\omega}\bar{w}^2 h_c\epsilon\int d^2x'\left(\tfrac{1}{2}(\nabla'z')^2 + \tfrac{1}{2}z'^2 - 16z'^3\right).\tag{4.40}$$

The instanton satisfies

$$\nabla'^2 z' = z' - \tfrac{1}{2}z'^2.\tag{4.41}$$

According to Eq. (4.39), the size of the nucleus is

$$Y_n \simeq \frac{\bar{w}^{1/2}}{(2h_c\epsilon)^{1/4}}\,\delta.\tag{4.42}$$

Tunneling via the formation of the nucleus will occur if the size of the defect, d, in the Y-direction exceeds Y_n, that is at

$$d > (w\delta)^{1/2}/(2h_c\epsilon)^{1/4}.\tag{4.43}$$

After the nucleus has been formed due to the tunneling, it expands in real time according to the classical equations of motion, Eq. (4.11).

The integral in Eq. (4.40), evaluated along the instanton trajectory, is a number of the order of unity. Consequently, the factor before the integral gives an estimate of the extremal action. In terms of the number of tunneling spins, $N = M_0 l Y_n \delta / \mu_B$, we obtain [42]

$$B_0 = 2^{1/4} k \frac{\gamma H_c}{\omega} \left(\frac{w}{\delta}\right)^{3/2} h_c^{1/4} \epsilon^{5/4} N, \tag{4.44}$$

where $k \simeq 1$ is the value of the integral. Comparison of Eq. (4.44) with Eq. (4.33) shows that in terms of N the geometry of the tunneling has little effect on the dependence of B on the parameters of the system. It appears to be quite universal in the limit of a low barrier.

4.2.3   The crossover temperature

In the absence of dissipation the temperature of the crossover from the thermal to the quantum regime is given by $k_B T_c = U_0 / B_0$. Therefore, one must calculate the height of the energy barrier, U_0. As has been shown above, the limit of a small barrier automatically provides the condition of small derivatives, $|\nabla z| \ll 1$. In this limit the energy of the static configuration of the wall is given by an equation similar to Eq. (4.24),

$$E = \sigma \delta^2 \int d^2x \left(\tfrac{1}{2}(\nabla \bar{z})^2 + u(\bar{z}) - h\bar{z}\right), \tag{4.45}$$

where $d^2x = dx_1 \, dx_2$ and $\nabla = (\partial_1, \partial_2)$. For a flat wall in $1+1$ dimensions $\nabla \bar{z} = 0$ and the height of the barrier is

$$U_0 = \sigma A_w u_0, \tag{4.46}$$

where u_0 is given by Eq. (4.29) at $\bar{z} = z_0$; $u_0 = \tfrac{2}{3} \bar{w}(2h_c \epsilon)^{3/2}$. In terms of the number of tunneling spins one obtains

$$U_0 = \frac{8\sqrt{2}}{3} \mu_B H_c \frac{w}{\delta} h_c \epsilon^{3/2} N. \tag{4.47}$$

Together with Eq. (4.33) this gives [42]

$$T_c = \frac{5\sqrt{2}}{36} \hbar \omega \left(\frac{\delta}{w}\right)^{1/2} h_c^{1/4} \epsilon^{1/4}. \tag{4.48}$$

Note that $T_c \simeq \hbar \omega_0$, where ω_0 is the instanton frequency of Eq. (4.37).

In $2+1$ dimensions the energy of the nucleus is

$$E = \sigma l \delta \int dx_2 \left[\tfrac{1}{2}\left(\frac{d\bar{z}}{dx_2}\right)^2 + u(\bar{z}) - h\bar{z}\right]. \tag{4.49}$$

A dimensional analysis shows that, for a small barrier, Eq. (4.49) can be presented as

$$E = \sigma l \delta \bar{w}^{3/2} (2h_c \epsilon)^{5/4} \int dy' \left[\frac{1}{2} \left(\frac{dz'}{dy'} \right)^2 + \frac{1}{2} z'^2 - 16z'^3 \right], \qquad (4.50)$$

where the integral is of the order of unity if it is taken along the extremal trajectory. The latter satisfies

$$\frac{d^2 z'}{dy'^2} = z' - \frac{1}{2} z'^2. \qquad (4.51)$$

The solution of this equation,

$$z = \frac{3w(2h_c \epsilon)^{1/2}}{\cosh^2 (2y/Y_n)}, \qquad (4.52)$$

gives the shape of the critical nucleus shown in Fig. 4.4. Further integration of Eq. (4.50) gives

$$U_0 = \frac{6}{5} \sigma l \delta \bar{w}^{3/2} (2h_c \epsilon)^{5/4} \qquad (4.53)$$

or the equivalent expression

$$U_0 = \frac{24\sqrt{2}}{5} \mu_B H_c \frac{w}{\delta} h_c^{1/2} \epsilon^{3/2} N, \qquad (4.54)$$

where $N = M_0 l Y_n \delta / \mu_B$ is the total number of spins inside the nucleus. On comparing Eq. (4.54) with Eq. (4.47) one finds that, in terms of N, the dependence of the barrier on the parameters is the same for different geometries of tunneling.

Equations (4.44) and (4.54) allow one to obtain the crossover temperature for the nucleation process [42],

$$T_c = \frac{12 \times 2^{1/4}}{5k} \hbar\omega \left(\frac{\delta}{w} \right)^{1/2} h_c^{1/4} \epsilon^{1/4}. \qquad (4.55)$$

Comparison with Eq. (4.48) again demonstrates that there is a universal dependence of T_c on the parameters. The only model-dependent parameter here is the width of the barrier, which one can associate with the size of the defect. Other parameters can be found from independent experiments.

4.2.4 The pre-exponential factor

Let us now estimate the prefactor A in the expression for the tunneling rate, $P = A \exp (-B_0)$. It is determined by the contribution to the path integral of tunneling trajectories that are small perturbations of the instanton. According to Chapter 2, the general expression for A in the case of a flat wall tunneling through a planar defect is

$$A = \left(\frac{B_0}{2\pi} \right)^{1/2} \left(\frac{\mathrm{Det}\, \hat{K}_0}{\mathrm{Det}'\, \hat{K}} \right)^{1/2}. \qquad (4.56)$$

Here Det$'$ means exclusion of the zero mode corresponding to the translational time-invariance of the instanton,

$$\hat{K}_0 = -\partial_\tau^2 + \omega_1^2$$
$$\hat{K} = -\partial_\tau^2 + \omega^2 u''(\bar{z}_i), \qquad (4.57)$$

where $\omega_1 = \omega_0/2$ is the frequency of small oscillations near the minimum of the potential at $\bar{z} = z_1$ (Fig. 4.5a), ω_0 is the instanton frequency given by Eq. (4.37), and $\bar{z}_i = z_2/\cosh^2(\omega_0\tau)$ is the instanton, Eq. (4.36).

The computation of A is, in general, a difficult problem. It can be done rigorously, however, in the case of a small barrier, which is of primary interest here. With the help of Eqs. (4.29) and (4.36) we obtain

$$u''(z_i) = \frac{1}{4}\left(1 - \frac{3}{\cosh^2(\omega_0\tau)}\right). \qquad (4.58)$$

Each determinant in Eq. (4.56) is the product of eigenvalues of operators in Eq. (4.57). The problem, therefore, reduces to the solution of

$$\left(-\partial_\tau^2 + \omega_1^2\right)\phi_n = \lambda_n\phi_n$$
$$\left[-\partial_\tau^2 + \omega_1^2\left(1 - \frac{3}{\cosh^2(\omega_0\tau)}\right)\right]\psi_n = \mu_n\psi_n. \qquad (4.59)$$

Exclusion of the zero mode, $\mu_n = 0$, leads to the proportionality of the ratio of determinants in Eq. (4.56) to ω_0^2. By evaluating this ratio one obtains

$$\frac{\text{Det}\left(-\partial_\tau^2 + \omega_1^2\right)}{\text{Det}'\left\{-\partial_\tau^2 + \omega_1^2[1 - 3/\cosh^2(\omega_0\tau)]\right\}} = \frac{\prod_n \lambda_n}{\prod_n' \mu_n} = 15\omega_0^2. \qquad (4.60)$$

This gives for the prefactor [42]

$$A = \left(\frac{15}{2\pi}\right)^{1/2} B_0^{1/2}\omega_0, \qquad (4.61)$$

where B_0 and ω_0 are given by Eq. (4.33) and Eq. (4.37), respectively.

Observation of tunneling requires a not very large B. Consequently, the value of the prefactor A for any reasonable experiment will be somewhere between ω_0 and $10\omega_0$. On the basis of dimensional arguments, one can see that this must be the case for any tunneling geometry, including the nucleation process studied above.

4.2.5   Additional thoughts on quantum depinning of domain walls

An experiment that would study tunneling of a single-domain wall through a planar defect may be somewhat similar to experiments on tunneling in Josephson junctions [52]. In the magnetic case the junction can be made of a layer of the material that is different from the bulk. The material of the junction should be selected such that it pins the wall. Modern evaporation techniques allow one to obtain a thickness of the junction (defect) as small as one atomic monolayer. The critical field H_c will, in general, depend on the thickness w and the material of the junction. The width of the potential well, which is produced by the junction itself, can hardly be significantly less than the domain wall width. Correspondingly, one should expect the parameter $\bar{w}$ in our formulas to be of the order of unity for $w < \delta$, and to be roughly given by w/δ for $w > \delta$. For a narrow junction, from Eqs. (4.55) and (4.44), we have

$$T_c \simeq \hbar\omega(h_c\epsilon)^{1/4} \tag{4.62}$$

for the crossover temperature, and

$$A \simeq B^{1/2} T_c/\hbar \tag{4.63}$$

$$B \simeq \frac{\gamma H_a}{\omega} (h_c\epsilon)^{5/4} N \tag{4.64}$$

for the prefactor and the WKB exponent of the tunneling rate, with $\Gamma = A\exp(-B)$, $\omega = v_0/\delta$, $h_c = H_c/H_0$, and $\epsilon = 1 - H/H_c$.

The above formulas allow one to estimate the effect on the basis of the data on the coercive field H_c, the anisotropy field H_a, and the parameters of the domain wall, namely the limiting velocity v_0 and the thickness δ. Equation (4.62) shows that the dependence of T_c on H_c and ϵ is rather weak [53,42], and T_c is mostly determined by the frequency ω. In the limit of large transversal anisotropy $K_\perp$, Eq. (4.23) gives $\hbar\omega \simeq (4\mu_B/M_0)(K_\parallel K_\perp)^{1/2}$. To within a numerical factor and the power of ϵ this coincides with the crossover temperature obtained by the exact solution of the equations of micromagnetic theory for tunneling of magnetization in single-domain particles and quantum nucleation of magnetic bubbles. For a uniaxial ferromagnet, $K_\perp$ in Eq. (4.23) must be replaced by the magnetic dipole energy $2\pi M_0^2$. In the limit of a large uniaxial anisotropy, $K_\parallel \gg 2\pi M_0^2$, this gives $\hbar\omega = 4\pi\mu_B M_0$. This energy is of order 0.1–1 K for typical values of the magnetization, $M_0 \simeq 10^2$–10^3 emu cm^{-3}. Given that the parameters h_c and ϵ in Eq. (4.62) are raised to the power one quarter, this means that T_c for quantum depinning of domain walls certainly falls within the accessible experimental range.

Let us now estimate how many spins can participate in the tunneling process. For $H_c \simeq 10$ Oe and $H_a \simeq 10^4$ Oe, the tunneling exponent B of Eq. (4.64) is of the order of $10^{-3}\epsilon^{5/4}N$. This suggests that, even at $\epsilon \simeq 10^{-2}$, that is without very fine tunning of the magnetic field, as many as 10^7 spins may coherently tunnel through the energy barrier.

4.3   Tunneling of antiferromagnetic domain walls

In weak ferromagnets, a small non-compensation,

$$m = \chi_\perp [d \times l], \tag{4.65}$$

arises from canting of equal sublattices ($m_1 = m_2$). Here $l = (m_1 - m_2)/(2m_1)$ is the Néel vector and d is the vector directed along the principal axis of the crystal, whose length equals the Dzyaloshinsky field. This non-compensation is typically of the order of $10^{-3}m_1$, so that m simply follows the dynamics of l. One should remember, however, that m couples to the magnetic field. There are two problems of tunneling in bulk antiferromagnets similar to those studied in ferromagnets. The first is quantum nucleation of a bubble of the reversed magnetization in the magnetic field opposite to m. The second is quantum depinning of antiferromagnetic domain walls. To study these problems one should add exchange terms ($\partial_i \equiv \partial/\partial x_i$),

$$a(\partial_i m_1)^2 + a(\partial_i m_2)^2 + b(\partial_i m_1 \cdot \partial_i m_2), \tag{4.66}$$

and the Dzyaloshinsky term,

$$-d \cdot [m \times l], \tag{4.67}$$

to Eq. (3.54) and replace V by the integration over the volume. Then a derivation similar to that for antiferromagnetic particles confirms that, at $d \leq (K/\chi_\perp)^{1/2}$ (which is usually the case in weak ferromagnets), the antiferromagnetic quantum dynamics dominates. The solution of the nucleation problem is also very much along the lines of that for ferromagnets. The crossover from the quantum to the thermal regime occurs at a temperature that has the same dependence on the parameters as does that in the problem of a small antiferromagnetic (AFM) particle.

Let us now briefly discuss quantum depinning of AFM domain walls. It should exist in materials with low coercivity, in which pinning barriers are small, and also near the coercive field, when barriers are lowered by the magnetic field. In the previous section, we have demonstrated that, for ferromagnetic domain walls, the temperature of the crossover from thermal to quantum depinning, to within an order of magnitude, is universally determined by $\hbar v_0/\delta$. Here v_0 and δ are the limiting velocity and the thickness of the domain wall, respectively. The argument and the estimate also hold for AFM domain walls. In that case [54],

$$v_0 = \tfrac{1}{2}\gamma\left(\frac{A}{\chi_\perp}\right)^{1/2}, \quad \delta = \left(\frac{A}{K_\parallel}\right)^{1/2}, \tag{4.68}$$

where A is the exchange constant. Quite remarkably, this gives [55]

$$T_c \simeq \hbar\gamma \left(\frac{K_\parallel}{\chi_\perp}\right)^{1/2}, \tag{4.69}$$

that is, the same answer as for AFM grains. Consequently, quantum tunneling of domain walls should reveal itself at higher temperatures in antiferromagnets than it does in ferromagnets.

4.4 The effect of dissipation on tunneling of domain walls

The effect of the dissipation on the tunneling of domain walls can be easily estimated along the lines of the Caldeira–Leggett approach. When the dissipation is linear and Ohmic, this approach allows one to relate the imaginary-time dissipative dynamics to the macroscopic friction coefficient. This method is well suited to the problem of domain wall tunneling since the real-time dynamics of domain walls is commonly described in terms of linear mobility. The latter can be obtained from an independent macroscopic experiment.

The mobility of the domain wall with respect to the magnetic field, μ_H, is defined as

$$v = \mu_H H. \tag{4.70}$$

One can connect it to a more conventional mobility, μ_F, with respect to the force, F, on the unit area of the wall, $v = \mu_F F$. Since $F = 2M_0 H$ we get $\mu_H = 2M_0\mu_F$. According to Eq. (2.60), the characteristic frequency associated with the dissipation is $\eta = 1/(\mu_F m_0)$, where m_0 is the mass of the unit area of the wall, $m_0 = \sigma/v_0^2$. Then Eq. (2.85) provides a rough estimate of the effect of dissipation on tunneling,

$$B = B_0 \left(1 + \frac{M_0 v_0^2}{\mu_H \sigma \omega_0}\right). \tag{4.71}$$

For typical numbers $M_0 \simeq 10^2\,\mathrm{emu\,cm^{-3}}$, $v_0 \simeq 10^4\,\mathrm{cm\,s^{-1}}$, $\mu_H \simeq 10^2\,\mathrm{cm\,s^{-1}\,Oe^{-1}}$, $\sigma \simeq 1\,\mathrm{erg\,cm^{-2}}$, $\omega_0 \simeq 10^{10}\,\mathrm{s^{-1}}$, the contribution of the dissipation to the WKB exponent is about 1%. The relative contribution of the dissipation to the tunneling rate is of the order of

$$\beta \frac{\gamma\delta}{\mu_H} \left(\frac{H_c}{H_a}\right)^{1/4} \epsilon^{-1/4}, \tag{4.72}$$

where β is a factor determined by the structure of the magnetic anisotropy. (For a rhombic crystal $\beta = [(1 + K_\perp/K_\parallel)^{1/2} - 1]$.) Typically, this contribution is small. One may also think, however, about the possibility of testing the effect of dissipation on tunneling in an experiment with domain walls. According to Eq. (4.72) dissipation becomes noticeable in materials with very low mobility of domain walls. Of course, the extraction of the contribution of the dissipation from experimental data would require changing the mobility without a significant change in other parameters. This can be done, e.g., by changing the concentration of impurities.

Let us also briefly discuss some particular mechanisms of the dissipation. Tatara and Fukuyama [43] studied the interaction of the ferromagnetic domain wall with Stoner excitations in metals. They found that the relative effect of the dissipation on tunneling is proportional to

$$\exp\left(-\frac{\pi\lambda|p_+ - p_-|}{\hbar}\right), \tag{4.73}$$

where λ is the domain wall thickness and $p_{\pm}$ are the Fermi momenta of electrons with different spin projections. For conventional ferromagnetic metals, like Fe, Co, and Ni, the factor in the exponent is 100–1000, so that the effect is absent. Theoretically, the effect can be non-negligible in weak itinerant ferromagnets with small $|p_+ - p_-|$ or in strong itinerant ferromagnets with very narrow domain walls, like, e.g., $SmCo_5$. Tatara and Fukuyama also found that, in typical metals, the dissipation due to eddy currents is small. In magnetic insulators the main source of dissipation at $T \simeq T_c$ should be magnons [53,42]. These excitations have a gap that is typically of the order of the instanton energy, $\hbar\omega_0$. The latter energy also provides the scale of the crossover temperature. Thus, at $T \ll T_c$, the effect on tunneling of the interaction of the domain walls with magnons should be exponentially small, and the main sources of dissipation should be impurities and phonons.

4.5   Tunneling of flux lines in superconductors

In recent years quantum tunneling of the magnetic flux lines in superconductors has received as much attention as has macroscopic quantum tunneling in magnets. Although it is not possible to cover both topics in this book, we decided to touch upon superconductivity because many experiments on tunneling of flux lines, as well as the data, are very similar to the experiments and the data on tunneling of magnetization. There is a certain physical similarity between the dynamics of a bulk ferromagnet due to the motion of domain walls and the dynamics of a superconductor due to the motion of flux lines. In both cases the slow dynamics is dominated by the decay of metastable states created by pinning of extended macroscopic objects, domain walls in the case of a ferromagnet and flux lines in the case of a superconductor. There is also an important difference between the two cases. Whereas the high-frequency motion of domain walls is mostly determined by their inertia, and to a lesser degree by dissipation, the motion of flux lines is believed to be almost entirely dissipative; their inertial mass can be neglected in most problems.

Most recent experiments on tunneling of flux lines have been done on high-temperature superconductors. In these materials, tunneling is much more pronounced because of the pancake structure of the flux line. The latter is due to the anisotropy of copper oxides. The effective mass of the electron, m_c, for the motion across CuO_2 planes, is large compared with the effective mass, m_{ab}, for the motion in the plane. In some materials, like, e.g., Tl-based high-T_c compounds, the ratio m_c/m_{ab} can be as large as 10^4. As a result, the flux line has a rather loose pancake structure of two-dimensional current rings (vortices) in consequent CuO_2 layers, coupled by weak Josephson and electromagnetic forces. The displacement of 2D vortices with respect to each other in that pancake structure up to to a distance $\lambda_J = (m_c/m_{ab})^{1/2}d$ (d being the interlayer distance) costs very little energy. Imagine now such a pancake structure on the background of the pinning potential formed by random positions of oxygen vacancies, Fig. 4.6. Vacancies, as well as other defects, pin the core of the vortex. There are typically a few oxygen vacancies within the core of a 2D vortex of radius $\xi \simeq 15$ Å. That radius determines the typical distance between the minima of the pinning potential. From that picture one should expect quantum diffusion of vortex lines in high-T_c materials to proceed via single events involving displacements of individual 2D vortices by a distance ξ. These are actually very light and small objects, so it is no wonder that quantum phenomena in high-temperature superconductors manifest themselves more strongly than they do in conventional superconductors.

Let us now apply the method of Chapter 2 to estimate the tunneling exponent for pancake vortices. The usual estimate of the WKB exponent is $(Um_{\text{eff}})^{1/2}\xi/\hbar$, where U is the magnitude of the pinning potential. The effective tunneling mass for the overdamped motion can be obtained from Eq. (2.82), $m_{\text{eff}} \simeq \eta/\omega$. On the other hand, from Newton's equation for the overdamped motion, $\eta\dot{x} = -dU/dx$, one has $U/\omega \simeq \eta\xi^2$. By combining these formulas and replacing

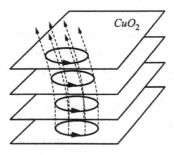

Figure 4.6 Pinning of pancake vortices in high-temperature superconductors.

η by the mobility of the 2D vortex, $\mu = 1/\eta$, we obtain the estimate for the friction-dominated tunneling:

$$B \simeq \frac{\xi^2}{\hbar\mu}, \tag{4.74}$$

where μ is the mobility of the vortex. The theoretical result for the mobility of the 2D vortex in a dirty superconductor is [56]

$$\mu = \frac{\xi^2}{\hbar}\left(\frac{\rho_n}{\rho_0}\right). \tag{4.75}$$

Here ρ_n is the normal sheet resistivity of the CuO_2 layer and ρ_0 is the quantum of the resistivity,

$$\rho_0 = \frac{\pi\hbar}{2e^2} \simeq 6.45\,\mathrm{k\Omega}. \tag{4.76}$$

By substituting Eqs. (4.75) and (4.76) into Eq. (4.74) we obtain [57] a simple estimate of the tunneling exponent,

$$B \simeq \frac{\rho_0}{\rho_n}. \tag{4.77}$$

At $\rho_n \simeq 200\,\Omega$, which is typical, this gives $B \simeq 30$, a reasonable value to ensure that tunneling of pancake vortices is observable. The temperature of the crossover from the thermal to the quantum regime is given by U/B, as usual. The value of the pinning barrier follows from the critical current measurements. At $U \simeq 150\,\mathrm{K}$ and $B \simeq 30$ one obtains $T_c \simeq 5\,\mathrm{K}$.

The above estimates are valid when the motion of vortices is dominated by friction. However, theory [58] and experiment [59] indicate that, at a very low temperature, the superconductor may enter the so-called Hall regime in which the motion of vortices is dominated by the Magnus force. In that case one obtains [60]

$$B \simeq \left(\frac{m_{ab}}{m_c}\right)^{1/2}\nu$$
$$T_c \simeq \frac{\hbar^2}{2\pi\xi^2 m_{ab}}, \tag{4.78}$$

where ν is the number of electrons per core of the 2D vortex. This also gives reasonable values of B and T_c. Thus, independently of the dynamics, one should expect tunneling of the magnetic flux in high-temperature superconductors to be observable at temperatures of a few kelvins.

The thermal overbarrier relaxation of flux line lattices has been the subject of intensive studies since the 1960s. The old Anderson–Kim model [61], as well as more recent vortex glass models [62], predict that the relaxation of the current and the magnetic moment due to the diffusion of flux lines in the random pinning potential must depend on $\ln(t)$ and some power of the temperature. As in the case of ferromagnets, quantum diffusion of flux lines must result in temperature-independent magnetic relaxation in the limit of a low temperature. We shall return to superconductors in Chapter 5, when discussing low-temperature magnetic relaxation.

Chapter 5

Quantum magnetic relaxation

5.1   The physics of slow relaxation in solids

At a given temperature any magnetic system has a certain state that corresponds to the absolute minimum of its free energy. For a bulk ferromagnetic crystal it is a certain configuration of domains. For a system of interacting monodomain particles, it is a certain orientation of individual magnetic moments. For a superconductor it is a certain structure of the flux-line lattice, etc. In practice, however, the minimum-energy state is difficult to achieve because there are various metastable states. In spin glasses this situation is generic: the complexity of the potential makes the energy minimum physically unattainable even in a microscopic volume. In contrast, in conventional magnetic crystals the spin structure on the scale of 100–1000 Å is usually the one that minimizes local interactions. On bigger scales, however, even well-ordered systems begin to exhibit metastability. This is because grain boundaries, dislocations, impurities, etc., do not allow domain walls to move freely in order to establish the configuration of domains that corresponds to the true energy minimum. Pinning of flux lines produces a similar effect in superconductors. A separate case is a system of monodomain particles embedded within a non-magnetic solid matrix. Here each particle can be in a metastable state due to energy barriers produced by the magnetic anisotropy.

The presence of metastable states results in magnetic hysteresis. All magnetic systems that exhibit hysteresis are expected to relax slowly toward the minimum of the free energy. Note that metastability is common among non-magnetic solids as well. For any given solid under stress there is a certain configuration of grain boundaries, dislocations, and impurities that corresponds to the absolute minimum of the free energy. The slow irreversible evolution of a solid toward that minimum may take hundreds of years. This phenomenon is know as *aging* of materials. It is physically very different from the fast relaxation toward the local thermodynamic equilibrium. An example of the latter is the relaxation of local elastic deformations caused by a suddenly applied mechanical stress. This relaxation, being carried out by phonons, takes very little time. A similar fast relaxation of the magnetization after, e.g., a change in the applied field, is carried out by spin waves. In all cases the fast relaxation can be understood in terms of damped harmonic oscillations near thermodynamic equilibrium. It falls, therefore, within the well-studied domain of linear-response theory and linear thermodynamics. In contrast, the slow relaxation toward the global thermodynamic equilibrium is a non-linear effect. Its long duration is due to the exponentially large lifetimes of metastable states. Consequently, it may be years before the decay of metastable states adds up to a noticeable change in a macroscopic parameter such as, e.g., the global magnetic moment of a solid. Progress in experimental techniques has made it possible to study the slow relaxation in virtually all materials. One universal feature uncovered by these measurements is the $\ln(t)$ character of the slow relaxation. A closely related phenomenon is the $1/f$ noise in the limit of low frequencies. Although no general rigorous theory of this phenomenon exists, it can be easily understood within the framework of simple, physically appealing, models. We will introduce them in the next section. Before doing so, let us discuss in more detail the physical picture of the slow relaxation.

At a high enough temperature the relaxation proceeds via local overbarrier transitions driven by thermal fluctuations. An example is the decay of the total magnetic moment of a permanent magnet, shown schematically in Fig. 5.1. In that example a uniform magnetization is obtained by applying a strong magnetic field, Fig. 5.1(a). When the field is removed, domains of opposite magnetization nucleate at the surface of the sample, Fig. 5.1(b). The actual process of domain formation is quite complicated. What is important is that, for a sufficiently large magnet, the minimum of the magnetostatic energy requires zero total magnetic moment, Fig. 5.1(d). An ideal ferromagnet would relax to that state almost immediately. In a real system, however, the initial fast propagation of domain walls is stopped by defects at some intermediate state with a non-zero total moment, Fig. 5.1(c). This is a *critical state* in which the local driving force on the domain wall is balanced by the local pinning potential. The critical state then slowly relaxes toward the absolute energy minimum. This occurs via rare hopping of domain walls from one trap to another. If the external field is now changed, domain walls expand like elastic membranes until a new critical state has been formed. The universality of the slow relaxation is due to the fact that it always begins in the critical state, in which energy barriers are just starting to

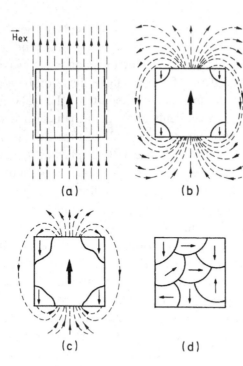

Figure 5.1 Relaxation of the magnetization in a permanent magnet: (a) the fully magnetized state, (b) initial fast propagation of domains of opposite magnetization, (c) the critical state from which the slow relaxation starts, (d) the final state of zero total magnetic moment.

develop. At any moment of time t that follows, the barriers which matter are the ones for which the lifetime is of the order of the running time t.

It is easy to see that a very similar situation occurs in a much simpler system of non-interacting single-domain particles. The relaxation time for an individual particle is proportional to $\exp(U/T)$, where U is the energy barrier. Owing to the different sizes and shapes of the particles, there is always some distribution of barrier heights. Thus, also in this case, when one is observing the system, the particles which are currently relaxing are the ones whose lifetimes coincide with the running time of the experiment t. We shall see that, for a macroscopic variable that is a sum of its local values, such as the total magnetic moment, this is enough to yield the $\ln(t)$ relaxation. By no means can this slow relaxation be confused with the fast evolution of the system toward local thermodynamic equilibrium. The latter is driven by dynamical equations, such as, e.g., the Landau–Lifshitz equation for the magnetic moment, whereas the slow relaxation is due to rare decays of local metastable states. It is important to understand that, no matter how complicated the interactions inside the system may be, no dynamics other than the decay of metastable states can explain relaxation processes in solids that last for days or years. When studying this process, an experimentalist is usually concerned with whether the relevant lifetimes fall within the experimental window. Indeed, owing to the exponential dependence of the lifetime on the temperature and the height of the barrier, even a moderate variation of parameters can bring the lifetime from microseconds to the age of the universe. It is, therefore, quite fortunate that the wide distribution of energy barriers in solids provides us with metastable states whose lifetimes appear within any experimental window. Of course, the fraction of such states may be small. Correspondingly, their decay may result in very little change in the measured macroscopic characteristic of the sample. Recent progress in the accuracy of measuring techniques has made it possible to detect such a change in materials that previously had been considered non-relaxing. It also made it possible to study the slow relaxation down to very low temperatures, at which quantum tunneling out of metastable states may dominate over thermal transitions. In this case one should expect relaxation that becomes independent of the temperature as one approaches absolute zero. During the last few years such a relaxation has, indeed, been observed in a variety of magnetic systems: small particles, ferromagnetic crystals, random magnets, superconductors, etc. In this chapter we shall concentrate on the relaxation experiments and their interpretation in great detail.

5.2   Models of slow relaxation

5.2.1   The critical state

The statistical mechanics of the critical state has yet to be developed. We will suggest a crude picture, however, which helps one to understand the $\ln(t)$ relaxation. Consider the time evolution of the total magnetic moment, M, of the system shown in Fig. 5.1. After the domain walls have been stopped by pinning centers, the further decay of M is due to domain-wall hopping in the presence of potential barriers. These potential barriers are formed by the pinning centers and the demagnetizing field (shown outside the body) that drives the walls inside. As the walls proceed inside the body the demagnetizing field formed by M decreases and so does the driving force on the walls. The fast initial motion of domain walls into the sample (the evolution from Fig. 5.1(b) to Fig. 5.1(c) stops at some $M = M_c$ when the demagnetizing field drops below the critical value at which barriers due to the pinning begin to develop. The simplest assumption for the evolution of barriers is that, for M dropping below M_c, they grow according to

$$U = U_0\left(1 - \frac{M}{M_c}\right). \tag{5.1}$$

Let us think for the moment that only thermal hopping of domain walls is involved. Then we can write that

$$\frac{\mathrm{d}M}{\mathrm{d}t} \propto \exp\left[-\left(\frac{U_0}{T}\right)\left(1 - \frac{M}{M_c}\right)\right]. \tag{5.2}$$

The solution of this equation is

$$M(t) = M(t_0)\left[1 - \frac{T}{U_0}\ln\left(\frac{t}{t_0}\right)\right], \tag{5.3}$$

where $M(t_0) \approx M_c$. The validity of the expansion (5.1) around M_c requires that $M(t)$ of Eq. (5.3) does not deviate much from M_c; that is, the system does not differ greatly from the critical state. Because of the logarithmic time dependence, this is always the case for $T \ll U_0$, that is, for a low enough temperature.

The coefficient in front of $\ln(t)$ is called the *magnetic viscosity*. We shall denote it in this chapter by the capital letter S,

$$S = -\frac{1}{M(t_0)}\frac{\partial M}{\partial \ln(t)}. \tag{5.4}$$

Our model predicts that, at $T \ll U_0$, $S = T/U_0$. The $\ln(t)$ relaxation and the proportionality of the viscosity to the temperature have indeed been observed in many systems (see below). At first sight this may look suspicious. Although real systems are believed to be in a critical state, it is hard to believe that the latter can be described in terms of one global parameter M and one barrier height (5.1). Rather, the driving and pinning forces on domain walls are balanced locally. Apparently, the success of our crude model is due to the fact that it catches the most important aspect of the slow relaxation: the barriers grow as the system is progressing toward the true equilibrium. Initially this model was suggested to explain the $\ln(t)$ relaxation of the magnetic flux (electric current) in type II superconductors [61]. In this case one should study the dynamics of flux lines instead of the dynamics of domain walls. All other aspects are the same: the (Lorentz) driving force, pinning, critical state, growth of barriers as the current drops below the critical value j_c, etc. The corresponding formula for the relaxation of the superconducting current is

$$j(t) = j(t_0)\left[1 - \frac{T}{U_0}\ln\left(\frac{t}{t_0}\right)\right], \tag{5.5}$$

where $j(t_0) \approx j_c$.

Our model predicts that, as the temperature decreases, the thermal viscosity must go to zero. Qualitatively, this follows from the fact that the system freezes in its metastable states. This behavior of the viscosity at moderately low temperatures $(T \ll U_0)$ has indeed been observed in many magnets and superconductors. However, deviations from that behavior have also commonly been observed in many systems at lower temperatures. It has been demonstrated experimentally that, below a few kelvins, many magnetic materials and the majority of high-temperature superconductors switch to a different relaxation regime, such that the relaxation is still logarithmic in time but the viscosity becomes independent of the temperature. Thus, a non-thermal relaxation is apparently taking place in these materials. At present no explanation for this phenomenon exists other than quantum tunneling of the magnetization in magnets and tunneling of flux lines in superconductors. Assuming this to be true, the temperature in Eqs. (5.3) and (5.5) must be replaced by the escape temperature, $T_{\mathrm{esc}}(T)$, of Chapter 2. Thus, measurements of the temperature dependence of the magnetic viscosity may shed light on the crossover from thermal to quantum decay of metastable magnetic states.

The above arguments, even though they are not rigorous, are physically appealing. A rigorous study of the magnetic tunneling requires the precise manufacturing of a well-characterized mesoscopic system such as, e.g., a single magnetic particle, or a single domain wall pinned by a single defect. In addition, an effort should be made to prepare the system in a state with a very low barrier. Although a certain amount of progress has been made in that direction, it remains a difficult experimental task. Bulk materials having many domain walls pinned by a variety of defects and particulate media with a distribution of particle sizes have the disadvantage of being characterized purely in terms of energy barriers, tunneling geometries, etc. On the other hand, however, there is a certain advantage associated with such materials. As has been explained above, they automatically appear in the critical state with low local barriers needed for tunneling. Consequently, the observation of temperature-independent slow relaxation in such materials should be a strong argument in favor of quantum tunneling.

5.2.2   Ensembles of small particles

Consider for a moment a system of identical non-interacting single-domain particles, frozen in a non-magnetic solid matrix, whose easy axes are parallel to each other and to the external magnetic field. Let M_+ and M_- be the total moments due to the particles whose individual moments are aligned along and against the direction of the field, respectively. They satisfy the following differential equations:

$$\frac{\mathrm{d}M_+}{\mathrm{d}t} = -\Gamma_+ M_+ + \Gamma_- M_-$$
$$\frac{\mathrm{d}M_-}{\mathrm{d}t} = -\Gamma_- M_- + \Gamma_+ M_+, \tag{5.6}$$

where $\Gamma_\pm$ are the rates of the transition out of the corresponding states. If the transitions are thermal, $\Gamma_\pm = \nu_\pm \exp\left[-K_\pm(H)V/T\right]$. For uniaxial particles whose axes are aligned with the field, $K_\pm = K(1 \pm H/H_{\mathrm{a}})^2$. On introducing the total moment of the system $M = M_+ - M_-$, one obtains

$$M(t) = M_{\mathrm{eq}} + (M_0 - M_{\mathrm{eq}})\exp\left(-\Gamma t\right), \tag{5.7}$$

where

$$M_{eq}(H) = \frac{\Gamma_- - \Gamma_+}{\Gamma_- + \Gamma_+} M_s \qquad (5.8)$$

is the equilibrium moment, $M_s = M_+ + M_- = $ constant is the total moment at saturation, $M_0 = M(t = 0)$, and $\Gamma = \Gamma_- + \Gamma_+$.

Consider now a system of uniformly oriented non-interacting uniaxial particles with the same magnetization and anisotropy energy per unit volume, but with the distribution of volumes, $f(V)$. Equation (5.7) must then be replaced by

$$M(t) = M_{eq} + (M(0) - M_{eq}) \frac{\int_0^\infty \mathrm{d}V f(V) V \exp\left[-\Gamma(V)t\right]}{\int_0^\infty \mathrm{d}V f(V) V}, \qquad (5.9)$$

where for thermal relaxation

$$\Gamma = \nu_- \exp\left[-K_-(H)V/T\right] + \nu_+ \exp\left[-K_+(H)V/T\right]. \qquad (5.10)$$

Typical values of νt in a macroscopic relaxation experiment are $\nu t \simeq 10^{11}-10^{15}$. For that reason the exponential factor under the integral in Eq. (5.9) is essentially the theta-function of $V - V_B$ (that is the function that is zero at $V < V_B$ and unity at $V > V_B$), where V_B is given by

$$V_B = \frac{T}{K_-(H)} \ln\left(\nu_- t\right). \qquad (5.11)$$

On taking account of this fact one obtains

$$M(t) = M(t_0) - [M(t_0) - M_{eq}(H)] \frac{\int_0^{V_B} \mathrm{d}V f(V) V}{\int_0^\infty \mathrm{d}V f(V) V}, \qquad (5.12)$$

where we have used that $\int_{V_B}^\infty = \int_0^\infty - \int_0^{V_B}$. The volume V_B can be called the blocking volume. At a time t only particles with volumes $V_B(t)$ effectively contribute to the relaxation. Smaller particles have already relaxed, whereas bigger particles have their moments blocked in the initial direction.

In the case of randomly oriented easy axes, the transition rates and the contribution of each particle to the total moment depend on the orientation of the particle with respect to the field. Thus, in general, the ensemble average must also contain averaging over the angles. To simplify the problem we shall assume that the field is small compared with the anisotropy field, as is usually the case in relaxation experiments. In this case the dependence of the energy barrier on the field and the orientation of the particle can be neglected, and the above equations remain valid.

If $f(V)V$ drops rapidly above a certain value, V_0, one can characterize the system in terms of the blocking temperature,

$$T_B = \frac{KV_0}{\ln\left(\nu t\right)}, \qquad (5.13)$$

where t is the characteristic measurement time of the instrument. On that time-scale, the transitions in most of the particles unfreeze above T_B. It should be emphasized that T_B is a characteristic of a state far from thermodynamic equilibrium. Its value is not absolute but rather depends on the type of measurement. For, e.g., Mössbauer studies of the relaxation, t is of the order of a few nanoseconds and T_B is significantly larger than that in static magnetization measurements. The condition $T \ll T_B$ (i.e., well below blocking) is mathematically equivalent to the condition $V_B \ll V_0$. In this case only a small part of the magnetization, namely the magnetization confined within particles of volume V_B, is relaxing; the magnetization of bigger particles is frozen.

A remarkable feature of the thermal relaxation follows from Eq. (5.12): *In the case of thermal relaxation, M depends on time only through the combination $T \ln (\nu t)$.* The derivation of Eq. (5.12), on which these conclusions are based, relies upon the assumption that $f(V)$ changes slower than exponentially in the vicinity of V_B. For typical statistical distributions, such as the Gaussian, Maxwellian, and exponential ones, this assumption is correct only for $V_B \ll V_0$ and becomes invalid at $V_B \geq V_0$. Thus, the $\ln (t)$ relaxation should be expected at $T \ll T_B$ but must cease at $T \geq T_B$. We shall see that this conclusion is supported by the experimental data.

According to Eq. (5.12), in the case of non-interacting particles at $H \ll H_a$, the definition of the magnetic viscosity, which is independent from the initial and final states, is

$$S = -\frac{1}{M_0 - M_{eq}}\frac{dM}{d \ln (t)} = \frac{T}{K\langle V \rangle}\frac{V_B f(V_B)}{\int_0^\infty dV f(V)}, \tag{5.14}$$

where $\langle V \rangle = \int_0^\infty dV f(V) V / \int_0^\infty dV f(V)$. The time-independent $S \propto T$ should occur only if, for small volumes, $f \propto 1/V$. In general, S must depend logarithmically on time. If $f(0) \neq 0$, the low-temperature viscosity of non-interacting particles must be proportional to $T^2 \ln (\nu t)$. Note also that, on the basis of Eq. (5.14), one may extract the distribution function from the relaxation measurement.

For cases in which the particles interact, the conception of the critical state may be more useful than are the formulas derived above. At first sight one might be puzzled by the fact that such physically and mathematically different models as the critical state and the barrier distribution lead to a relaxation that is logarithmic in time. The explanation lies in the fact that both models have the same basic feature: as the observation time is running, the system arrives at greater and greater barriers that are more and more difficult to overcome. This is a very general feature of complex systems (not necessarily magnetic systems) and the key to understanding the $\ln (t)$ relaxation. The most important consequence of the above consideration for the subject of this book is that the study of the temperature dependence of the relaxation can reveal the crossover from thermal to quantum relaxation. Indeed, in the case of purely thermal relaxation, all of the above formulas show that the relaxation should completely freeze as the temperature goes to zero. In the case of quantum relaxation, however, T in all formulas must be replaced by the escape temperature $T_{esc}(T)$, as has already been discussed for the critical state. Quantum transitions must reveal themselves in a non-zero, temperature-independent magnetic viscosity in the limit of $T \to 0$. They should also result in the breakdown of the scaling, $M = M(T \ln(\nu t))$, below the temperature of the crossover from the thermal to the quantum regime. We shall see that this is, indeed, the case for some magnetic systems for which tunneling is expected from theoretical models.

Finally, we would like to comment on the possibility that the thermal relaxation due to a certain distribution of energy barriers mimics the temperature-independent quantum relaxation. According to Eq. (5.12), the volume distribution $f(V) \simeq 1/V^2$ results in a relaxation that is proportional to $\ln[T(\ln(\nu t))]$. In an experiment this can be confused with the temperature-independent relaxation. In general, any distribution of energy barriers that is singular at $U \to 0$ poses the same danger. A singular distribution of barriers has been obtained [28] in numerical simulations of metastable spin configurations in ferrimagnetic grains with low exchange energy, and the question of whether the resulting weak temperature dependence of the relaxation could be confused with quantum relaxation was brought up. In our opinion, this is unlikely, because the $\ln(T)$ temperature dependence of the relaxation automatically implies its $\ln[\ln(t)]$ time dependence, which is likely to be perceived as no relaxation at all. An important additional check should be the thermal scaling plot, $M(T \ln(\nu t))$. Independently from the barrier distribution, quantum relaxation should result in a departure from the thermal scaling below the crossover temperature.

5.2.3   The zero-field-cooled magnetization curve

The independent source of information about the system formed by magnetic clusters of different sizes is the zero-field-cooled (ZFC) $M(T)$ curve. Consider a system of small particles with the volume distribution $f(V)$ cooled to a very low temperature at $H = 0$. The moments of the particles are then oriented randomly such that the total M is zero. Let us now apply a small field $H \ll H_a$ and study the temperature dependence of M. At a given temperature T, particles of volume $V < V_B$ are superparamagnetic, whereas particles of volume $V > V_B$ are blocked in the initial states. The moment along the field, induced in dN superparamagnetic particles of volume V, whose easy axes form the angle θ with the direction of the field, is

$$m_0 V \cos\theta \frac{\Gamma_- - \Gamma_+}{\Gamma_- + \Gamma_+} dN, \qquad (5.15)$$

where m_0 is the magnetization of the material of the particle. For a very weak field the rates can be written as $\nu \exp\left[(-KV \pm m_0 VH \cos\theta)/T\right]$ and Eq. (5.15) reduces to $\left[(m_0 V \cos\theta)^2 H/T\right] dN$. By introducing now the distribution function, $f(V)$, normalized according to $\int_0^\infty dV f(V) = N$, we obtain for the total ZFC moment

$$M(T, H, t) = \frac{m_0^2 H}{2T} \int_0^{V_B(T,t)} dV f(V) V^2, \qquad (5.16)$$

where the factor $\frac{1}{2}$ came from taking the average over the angle. At $T > T_B$, that is, at $V_B > V_0$, the integral in Eq. (5.16) becomes constant and M becomes proportional to $1/T$. This is the limit at which all particles are superparamagnetic. At low temperature, $T \ll T_B$, that is, at $V_B \ll V_0$, the temperature dependence of the total moment depends on the distribution function. For $f \propto V^n$, Eq. (5.16) gives $M \propto T^{2+n}$. Consequently, at low temperature, the ZFC magnetization grows with T for distributions with $n \geq -2$. Most magnetic systems with randomness exhibit an increase in the ZFC moment until T_B, followed by behavior according to the Curie law above T_B, in accordance with our discussion. At $n = -2$, which is the 'dangerous' distribution mentioned above, the ZFC magnetization curve must be constant below T_B. Such a distribution, therefore, must reveal itself in a very different ZFC magnetization curve. So should the distributions with $n < -2$, which should result in $M \propto 1/T$ throughout the entire temperature range. For such distributions, the particles which determine the bulk of the magnetization are always superparamagnetic so that T_B is undefined. The conventional experimental definition of T_B is the temperature at which the ZFC magnetization curve goes through the maximum. It is easy to obtain from Eq. (5.16) that T_B satisfies

$$V_B \frac{df}{dV_B} + 2f(V_B) = 0. \qquad (5.17)$$

For standard distributions that drop exponentially above a certain volume V_0, this gives T_B of Eq. (5.13) times a numerical factor of order unity that depends on the distribution. For, e.g., $f = \exp(-V/V_0)$ the experimental T_B is $2KV_0/\ln(\nu t)$, whereas for $f = \exp(-v^2/V_0^2)$ it is $KV_0/\ln(\nu t)$. It is interesting to notice that, although the ZFC magnetization and the viscosity are given by different formulas, Eq. (5.16) and Eq. (5.14) correspondingly, the maximum of the viscosity occurs at the temperature given by exactly the same equation, namely Eq. (5.17). Thus, for weakly interacting particles, the difference in the position of the maximum in the ZFC magnetization and viscosity measurements should be solely due to the difference in the measurement time. This may provide an independent test of the consistency of the experimental method.

In the presence of quantum relaxation, the temperature in the definition of V_B must be replaced by $T_{esc}(T)$. Thus, at $T < T_c$, quantum transitions, independently of the distribution, must result in a ZFC curve proportional to $1/T$ in the limit of low temperatures, Fig. 5.2. This would reflect the fact that some of the smallest particles remain superparamagnetic owing to quantum transitions down to absolute zero. It should be noted, however, that the contribution of these particles to the ZFC magnetization may be below the experimental threshold of detection. This is in contrast to the relaxation measurements, which are sensitive to the change in the total moment. The latter measurements, therefore, are better suited for the study of magnetic tunneling. Another reason that makes the relaxation studies more conclusive than are the ZFC studies is that the quantum superparamagnetic behavior shown by the dashed line in Fig. 5.2, if it were observed in an experiment, would be difficult to distinguish from the thermal contribution of volume distributions like those with the small-V behavior faster than $1/V^2$. In the relaxation measurements, however, both the temperature dependence and the time dependence of the moment are obtained. This allows one to check whether the $T \ln (\nu t)$ scaling of the relaxation predicted by the thermal theory holds at low temperatures.

5.2.4   Computer studies of slow relaxation

The physical arguments presented above can be illustrated by the computer results for the relaxation in an ensemble of single-domain particles for different distributions of barrier heights [63–65].

5.2.4.1 Non-interacting particles

The time dependence of the magnetization of a set of single-domain particles with volume V may be numerically computed [63,64] from Eq. (5.18):

$$\frac{\mathrm{d}M(V,t)}{\mathrm{d}t} = -\Gamma(V,t)\big(M(V,t) - M_{\mathrm{eq}}(V,t)\big), \tag{5.18}$$

where Γ is the rate of transitions and $M_{\mathrm{eq}}(V,t)$ is the equilibrium magnetization for given H and T at a time t. For a system composed of N particles with a size distribution $f(V)$, the time dependence of the magnetic moment can be obtained by integrating Eq. (5.19):

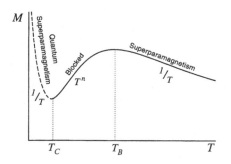

Figure 5.2 The ZFC magnetization curve: the solid line shows the typical behavior expected from Eq. (4.16), the dashed line shows a hypothetical low-temperature growth due to quantum superparamagnetism [64].

$$M(t) = \frac{1}{N} \int M(V,t) V f(V) \, dV. \tag{5.19}$$

Four different distributions $f(V)$ have been used to compute $M(t)$ using Eq. (5.19):

(i) identical particles: $f(V) = \delta(V - V_0)$
(ii) a uniform distribution: $f(V) = $ constant if $V < V_0$ and
   $f(V) = \exp(-V/V_0)$ if $V > V_0$.
(iii) $f(V) \propto 1/V$ if $V < V_0$ and $f(V) = \exp(-V/V_0)$ if $V > V_0$.
(iv) the log-normal distribution $f(V) \propto (V/V_0) \exp\left[-\alpha \ln^2(V/V_0)\right]$.

Fig. 5.3 shows the dependence of the total magnetic moment on time. The particles were initially magnetized to saturation, M_0, along the easy axis of anisotropy and then left free to relax toward the final state of zero magnetic moment in the absence of the field. We are particularly interested in the low-temperature regime, under which a small fraction of the total magnetic moment is relaxing. In the case of identical particles the decay of the total magnetic moment is exponential with time for all temperatures. For all other distributions the ln (t) relaxation is a good approximation, see Fig. 5.3. At larger times or higher temperatures, when $T \ln(t)$ becomes sufficiently large and the total M changes significantly, the $\ln(t)$ dependence of the relaxation is lost in accordance with our analytical model.

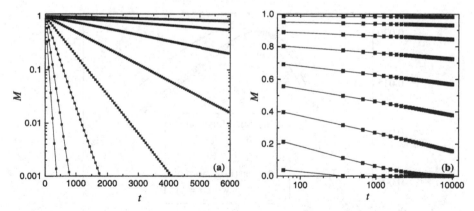

Figure 5.3 The time evolution of the total magnetic moment for different distributions of particle volume: (a) identical particles and (b) a log-normal distribution.

Figure 5.4 shows the temperature dependence of the magnetic viscosity for the above three ensembles ((ii), (iii), and (iv)) of non-identical particles. In the three cases the viscosity goes to zero at $T \to 0$ even though the law is different for each case. It has also been verified that the viscosity is constant for the distribution proportional to $1/V^2$, in agreement with the analytical prediction. When the temperature approaches the blocking temperature, the viscosity goes through a maximum and then decreases in accordance with our conclusion that $S(T)$ must reflect the barrier distribution. Above the blocking temperature the $\ln(t)$ relaxation is not a good approximation and the viscosity, S, depends on the time via $\ln(t)$. Its value for each T has been taken in the middle of the time interval. When the relaxation is studied in the presence of an external magnetic field (that is, after saturation the external field is switched to a non-zero value) the viscosity data are similar to those discussed above. In this case, the maximum of the viscosity shifts to lower temperature. These results are in agreement with those obtained from Monte Carlo simulations of the dynamic of an ensemble of 36 000 particles with different size distribution functions [65].

5.2.4.2 Interacting particles

The effect of the dipole–dipole interaction between the particles has been studied in Monte Carlo simulations [66]. A mean-field model was employed in which each individual moment experiences the demagnetizing field created by the total magnetization of the system. In contrast to the previous case, the barriers now

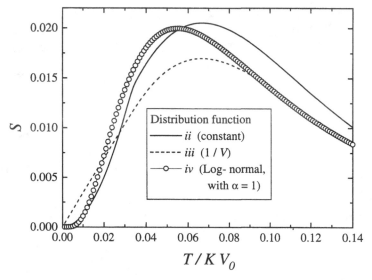

Figure 5.4 The magnetic viscosity, $S(T)$, computed for the distributions (ii), (iii), and (iv), described in the text.

change with time as the total moment of the system decreases. This model has features both of the critical state and of the barrier height distribution. The case of interacting particles is illustrated in Fig.5.5. For the weak interaction, Fig. 5.5(a), the relaxation law is still close to exponential with time. As the interaction increases, it switches to the ln (t) relaxation, Fig. 5.5(b). For other distribution functions the relaxation in the presence of the interaction is qualitatively the same as that without interaction. The tendency of the thermal magnetic viscosity to go to zero at $T \to 0$ remains unchanged by the interactions, as shown in Fig.5.6. We conclude, therefore, that neither analytical models nor computer

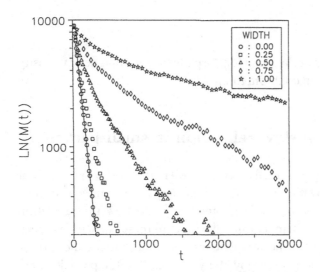

Figure 5.5 The time evolution of the total magnetic moment of an ensemble of 36 000 interacting particles for different widths of the size distribution: (a) weak interaction and (b) strong interaction. (Cited from [65,66].)

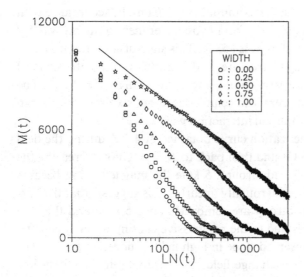

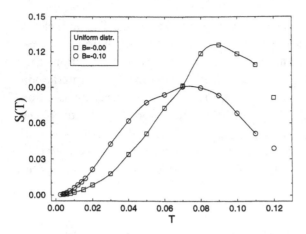

Figure 5.6 The temperature dependence of the magnetic viscosity for the cases of independent particles (squares) and dipole–dipole interaction between particles (circles). (Cited from [65,66].) The average dipole field B is in units of the anisotropy field.

simulations can provide any explanation for the plateau in the magnetic viscosity at low temperature other than quantum tunneling.

5.3   Experiments on slow relaxation in small particles

We shall start with an example of small (average size 4 nm) γ-Fe_2O_3 particles that do not exhibit tunneling down to 0.6 K. The purpose of this example is to show how the particles can be characterized magnetically and how the relaxation data can be analyzed. The low-temperature dynamics of these particles has been studied by Vincent *et al.* [67]. The sample consisted of a frozen colloid of γ-Fe_2O_3 particles. The electron-microscopy study showed that the particles were near spherical, with the diameter distributed around a peak value of 4.2 nm and the width of the distribution at half-maximum being 3.0 nm. In accordance with these data the mean diameter obtained from Mössbauer measurements was 5.2 nm. The Curie temperature of γ-Fe_2O_3 is 863 K. This suggests that thermal fluctuations of the ferrimagnetic spin structure of the particles at the atomic level did not play any role in the experimental temperature range, 0.6–12 K. The separation of the particles from center to center was about six times their mean diameter, suggesting that their mutual interaction was weak.

Figure 5.7 shows the magnetization curve obtained at 4.3 K during the field cycle from $H = 0$ to $H = 15$ kOe and then back to $H = 0$. Upon increasing the field, no real saturation was found; around 5 kOe the magnetization became reversible but continued to increase roughly linearly. This suggests that the barriers due to the magnetic anisotropy are removed above 5 kOe and that the increase in the total moment is due to the progressive canting of ferrimagnetic sublattices by the field. The observed slope in high field is indeed in accordance with the expected value of the exchange field. By extrapolating the high-field

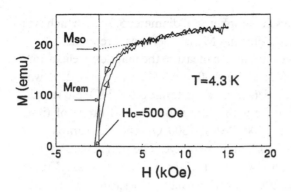

Figure 5.7 The magnetization curve of the ensemble of γ-Fe_2O_3 particles at 4.3 K. (Data from [67].)

behavior linearly to zero field, the authors of [67] determined the zero-field magnetization of the particles to be $M_{so} = 191$ emu. After they had magnetized the sample in 15 kOe, the remanent magnetization at zero field was $M_{rem} = 92$ emu $= 0.48 M_{so}$ per unit magnetic volume, which is close to the $0.5 M_{so}$ expected when all individual moments are oriented along their randomly distributed easy axis, in the direction selected by the field. The anisotropy energy was estimated from the magnetization curve to be 8×10^5 erg cm^{-3}, which is typical for this material.

Field-cooled (FC) and zero-field cooled (ZFC) magnetization curves are shown in Fig. 5.8. The residual magnetization, $M_0 = 94$ emu, remaining from the high-field (15 kOe) measurements, which is close to the previously obtained $M_{rem} = 92$ emu, has been subtracted from M to emphasize the effect of the 300 Oe field. The ZFC curve shows a broad peak around 13 K. This temperature corresponds to the blocking of particles of a typical volume V_B that satisfies $t = \tau_0 \exp (K V_B / T)$. On substituting here $K = 8 \times 10^5$ erg cm^{-3} and $T = 13$ K, and choosing $\ln (t/\tau_0) = 30$, which at $\tau_0 = 10^{-10}$ s corresponds to the ZFC experimental time scale $t \simeq 1000$ s, one obtains $V_B = 6.7 \times 10^{-20}$ cm^3. This

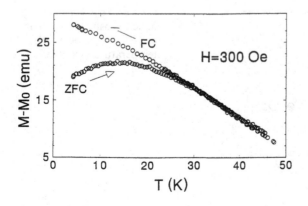

Figure 5.8 FC and ZFC magnetization curves of γ-Fe_2O_3 particles. (Data from [67].)

volume is equivalent to the particles being spheres of diameter 5.0 nm, which is in accordance with the electron-microscopy and Mössbauer measurements.

The following procedure has been applied to measure the magnetic relaxation. The sample was first heated to 50 K and a +300 Oe field was applied. It was then cooled to the measurement temperature (in the range 0.6–12 K). No relaxation was detected at that temperature within the experimental range of time. The field then was reversed from +300 Oe to −300 Oe and the sample was allowed to relax toward a new equilibrium at −300 Oe. The magnetization was recorded for a few hours, starting a few seconds after the field reversal. Figure 5.9 displays a few typical relaxation curves. For each temperature the average slope $dM/d\ln(t)$ had been determined by fitting a linear $\ln(t)$ behavior in the 5–6000 range. The slope was divided by the amplitude of the relaxation, $\Delta M_{FC} = M_{FC}(+300\,\text{Oe}) - M_{FC}(-300\,\text{Oe})$, which exhibited a 7% variation in the 0.6–12 K temperature range.

The magnetic viscosity is displayed in Fig. 5.10. Between 0.6 and 7 K a clear proportionality of the viscosity to the temperature was observed. The relationship extrapolated nicely to zero to within the experimental accuracy. These are

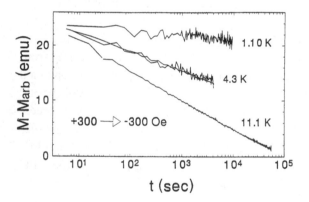

Figure 5.9 Typical relaxation curves obtained for γ-Fe$_2$O$_3$ particles after field reversal. (Data from [67].)

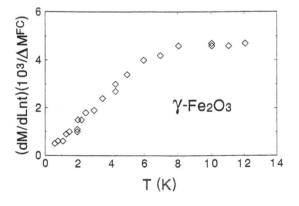

Figure 5.10 The magnetic viscosity for a system of γ-Fe$_2$O$_3$ particles versus temperature. (Data from [67].)

the characteristics of the purely thermal relaxation. The flattening of the viscosity above 7 K is, apparently, a vestige of the blocking at 13 K observed in the ZFC magnetization measurements.

As has been discussed in the previous section, one can obtain useful information by plotting the magnetization versus $T \ln (t/\tau_0)$. In the case of a purely thermal relaxation all data obtained at different temperatures must be assembled into a unique continuous curve. This was indeed achieved [67] for γ-Fe$_2$O$_3$ particles by choosing $\tau_0 = 10^{-10}$ s, Fig.5.11. The derivative of the master curve must represent the barrier-height distribution function. It can be seen from Fig. 5.11 that the master curve does indeed exhibit a slight curvature, varying from negative to positive for increasing values of the main barrier height, $T \ln (t/\tau_0)$. It can then be concluded from magnetic measurements that the distribution of particle sizes peaks at 3.7 nm (the inflection point of the master curve) and begins to decrease significantly between 4.3 nm (the saturation of the viscosity above 8 K) and 5 nm (the broad ZFC peak at around 13 K). This is in good agreement with the data from electron microscopy.

Very similar data on γ-Fe$_2$O$_3$ particles were also obtained in [68] both for the magnetic viscosity and for the scaling of the $T \ln(t)$ data until 1.8 K. These particles have a relatively low anisotropy field, less then 5 kOe, as can be deduced from the magnetization curve, Fig. 5.7. By using as a rough estimate of the crossover temperature $T_c \simeq \mu_B H_a$, we obtain that quantum relaxation in these particles should not be expected above 0.3 K. This may explain why the very-low-temperature data both for the ZFC magnetization and for the viscosity increase below 0.3 K have very recently been observed [69]. The above experiments and their analysis, however, show that our understanding of the thermal behavior of small particles is rather good and that there is a good chance of detecting deviations from that behavior due to quantum tunneling. We now turn to the data on small particles that do exhibit such a deviation.

A water colloid of ferrimagnetic CoFe$_2$O$_4$ particles of average diameter 3.5 nm was studied in [70]. Electron microscopy revealed that the particles were

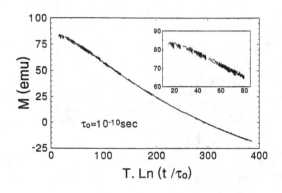

Figure 5.11 The magnetization of an ensemble of γ-Fe$_2$O$_3$ particles versus $T \ln (t/t_0)$. (Data from [67].)

nearly spherical and more or less uniformly distributed in size in the range 2–6 nm. The electron and X-ray diffraction patterns revealed the crystalline structure of the particles, the symmetry of the crystalline lattice being the same as that of the bulk $CoFe_2O_4$. The Curie temperature of this material is 793 K; no detectable change in the spin structure is expected at low temperature. The volume concentration of the particles in the colloid was about 10%, so that their dipole interaction was very weak compared with the anisotropy energy (see below). At 250 K the magnetization curve is nearly reversible, and the coercive field does not exceed a few oersteds. Hence, at this high temperature, the system is superparamagnetic. The dependence of the coercive field on the temperature in the range 2–250 K is displayed in Fig. 5.12, which shows that the number of particles that become superparamagnetic (unblocked) grows with increasing temperature. The fact that the coercive field does not saturate as the temperature is lowered indicates that blocking of the smallest particles persists down to the lowest experimental temperature, 1.8 K. The existence of such particles with low energy barriers makes this system interesting for the study of quantum tunneling.

The magnetic hysteresis at 2.4 K is shown in Fig. 5.13. The hysteresis loop closes at about 5 T, which roughly represents the highest anisotropy field inside the particles. Note that the bulk anisotropy of $CoFe_2O_4$ is 2×10^6 erg cm^{-3}, and the bulk magnetization is about 420 emu [71]. This gives the bulk anisotropy field $H_a = 2E_a/M_0$ of about 1 T. The experimental finding for particles is $H_a \approx 5$ T. The fact that the observed anisotropy field is large compared with the magnetization suggests that the anisotropy of the particles is most likely to

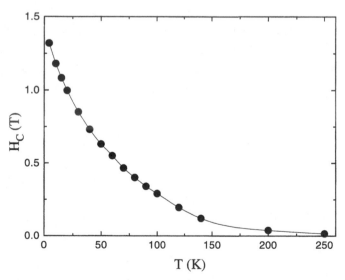

Figure 5.12   The temperature dependence of the coercive field for $CoFe_2O_4$ particles deduced from isothermal hysteresis loops. (Data from [70].)

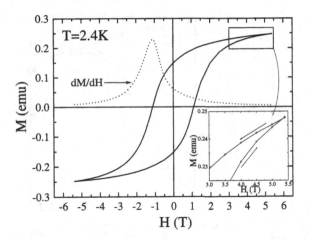

Figure 5.13 The magnetization versus applied field obtained at $T = 2.4$ K for $CoFe_2O_4$ particles. The inset is the high-field region indicates that the highest anisotropy field is larger than $5\,T$ [70].

be of crystalline origin rather than due to the shape, especially because the latter was found to be close to spherical. In any case the anisotropy field is greater by a factor of 10 than that for γ-Fe_2O_3 particles. This should shift the crossover from the thermal to the quantum regime into the kelvin range.

The FC and ZFC magnetization data at various fields are shown in Fig. 5.14. The ZFC curve has a broad maximum corresponding to the broad distribution of sizes. As expected, the maximum shifts to lower temperatures with increasing H because the field lowers the barriers. At low field the FC and ZFC curves merge at about 300 K. This gives the barrier height $U \approx 30T_B = 770$ meV. The typical volume of the particles responsible for the blocking can be estimated to be $V = U/E_a$. With the above numbers for U and E_a, the size of these particles is calculated to be about 7 nm, which is in accordance with the electron-microscopy data suggesting a uniform size distribution up to 6 nm. Note that, in this case, the largest particles determine the blocking temperature because of their dominating contribution to the total magnetization.

The magnetic relaxation measurements were performed down to 1.8 K. The system was cooled in +5 kOe from 300 K down to the measurement temperature, after which the field was rapidly reversed to −4 kOe. The variation of the total magnetization with time was recorded for a few hours. The $\ln(t)$ relaxation was observed for all temperatures below the blocking temperature, Fig. 5.15. It follows from Fig. 5.15 that the lower the temperature the smaller the fraction of the total magnetization that is relaxing during the observation time. This is in accordance with the expectation that the size of the particles which contribute to the relaxation decreases as the temperature is lowered, so that the contribution of the relaxing particles to the total magnetization decreases with temperature. Bigger particles become progressively blocked as the temperature decreases.

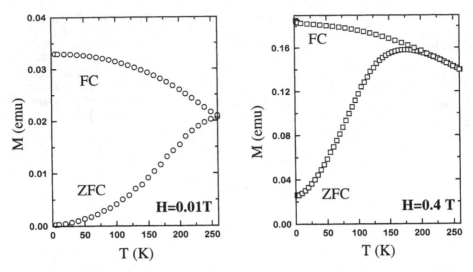

Figure 5.14   ZFC and FC magnetization curves obtained with various applied fields for $CoFe_2O_4$ [70].

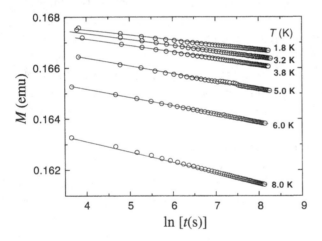

Figure 5.15   The decay with time of the magnetic moment versus $\ln(t)$ for various temperatures obtained for $CoFe_2O_4$ particles [70].

Note that, according to our estimate, the particles of size 7 nm were already blocked at 270 K. Assuming that T_B is proportional to the cube of the diameter, this means that particles contributing to the thermal relaxation at 3 K should be within the lower portion of the size distribution determined by electron microscopy, $d < 2$ nm.

The temperature dependence of the magnetic viscosity is shown in Fig. 5.16. The viscosity decreases monotonically by almost two orders of magnitude as the temperature decreases from T_B to 2.5 K. Although within this temperature range the actual dependence of S on T is more complicated

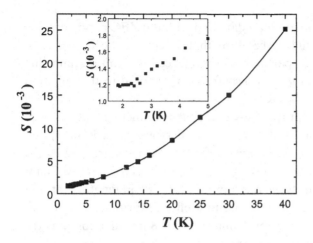

Figure 5.16 The magnetic viscosity, $S(T)$, extracted from the decay with time of the magnetic moment [70]. The inset represents the low-temperature data.

than a simple proportionality to T, there is an apparent change in the slope at 2.5 K. Below this temperature the viscosity becomes nearly independent of temperature. This could be an indication that quantum tunneling is occurring in the smallest particles present.

The existence of a non-thermal relaxation in $CoFe_2O_4$ at low temperature is also supported by the $T \ln (t/\tau_0)$ (with $\tau_0 = 10^{-10}$ s) plot shown in Fig.5.17. The relaxation data collected above 7 K assemble nicely into the universal curve expected in the case of purely thermal relaxation. Below 5 K there is an apparent systematic departure of the data from the universal curve. As has been discussed in the previous section, the anomaly in the magnetic relaxation could be due to a singular size distribution at $V \to 0$. To explain the observed anomaly in terms of there being such a distribution, one would have to assume the presence of a huge number of particles with sizes less than 2 nm. This is, however, in disagree-

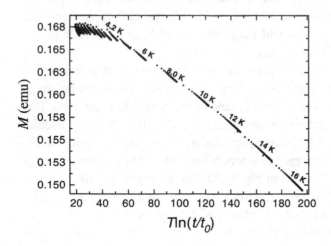

Figure 5.17 The magnetization of $CoFe_2O_4$ particles versus $T \ln (t/\tau_0)$.

ment both with the electron-microscopy data and with the ZFC magnetization data. The authors of [70] concluded, therefore, that the entire collection of magnetic measurements of $CoFe_2O_4$ particles gives evidence for quantum tunneling in the smallest particles, with the crossover from the thermal to the quantum regime occurring in the range 2.5–5 K.

Another interesting system for the study of magnetic tunneling that has been discussed intensively during the last few years is antiferromagnetic horse-spleen ferritin. In the context of MQT the interest in ferritin has been aroused by the work of Awschalom et al. [13] and its criticism by Anupam Garg [72] (see Chapter 6). Here we shall only address the question of tunneling in ferritin, not that of the coherence [13], which is a much more subtle effect. Ferritin is an iron-storage protein. It has a spherical cage of diameter about 8 nm that contains the mineral ferrihydrate combined with a phosphate. It has been characterized structurally by dark-field transmission electron microscopy [73], and by X-ray [74] and electron diffraction [75]. The core is equivalent to a small antiferromagnetic particle of hematite, α-Fe_2O_3. The fully packed ferritin contains about 4500 Fe^{3+} ions. The filling of the ferritin cage varies for different molecules, leading to a broad distribution of core sizes in natural ferritin in the range 3–7.5 nm [75,76]. Mössbauer studies [77,78] of ^{57}Fe-enriched horse-spleen ferritin show a well-defined magnetic sextet at 4.2 K that, as the temperature is raised, changes into a doublet within the temperature interval 10–15 K. This has been interpreted as superparamagnetic behavior at high temperature and a blocked magnetic state at low temperature, with the blocking temperature somewhere in the range 10–50 K, depending on the core size. Note that the characteristic time of Mössbauer measurements is 2.5×10^{-9} s, that is, 10 orders of magnitude smaller than that of DC magnetic measurements. For that reason (see the definition of the blocking temperature in Section 4.2) the Mössbauer blocking temperature should be significantly higher than T_B introduced earlier. The antiferromagnetic nature of the spin order in ferritin has been confirmed by observing the splitting of Mössbauer lines in the magnetic field [77] and by observing the spin-flop transition in high field [79]. The low-field magnetic susceptibility of ferritin [80] is also in agreement with its antiferromagnetic structure.

Ferritin molecules have small magnetic moments owing to the non-compensation of collinear spin sublattices arising from the finite size and irregular shape of the antiferromagnetic core. The FC and ZFC magnetization data [80] obtained from a natural ferritin sample containing 1.2×10^{16} protein molecules are shown in Fig. 5.18. Above 13 K the particles are superparamagnetic: the FC and ZFC curves merge. The magnetic susceptibility in that range of temperatures follows the Curie law (see the inset in Fig. 5.18). The average magnetic moment deduced from the Curie law corresponds to 15 non-compensated Fe^{3+} ions. Assuming that the main contribution to the total magnetization comes from the biggest particles, this is roughly in accordance with the number of non-compen-

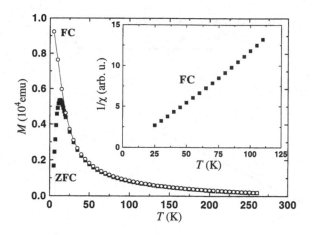

Figure 5.18 The temperature dependence of the magnetization for ferritin molecules obtained in the ZFC and FC processes with the applied field $H = 0.01\ T$. Inset: the inverse FC susceptibility versus temperature. (Data from [80].)

sated ions, $(4500)^{1/3}$, expected from the randomness of the core surface. Such a low moment means that the magnetic dipole interaction between the particles is very weak, which is also supported by the fact that the inverted susceptibility in the superparamagnetic regime extrapolates to zero at $T \to 0$. The low-temperature magnetic anisotropy of bulk hematite is $2.5 \times 10^5\ \mathrm{erg\ cm}^{-3}$ [71]. Assuming that the method works for ferritin particles, the volume of the particles responsible for the blocking can be found in the same manner as it was done in the previous two examples. For $T_B = 13$ K and $E_a = 2.5 \times 10^5\ \mathrm{erg\ cm}^{-3}$ one has particles of diameter 7.5 nm, in good agreement with the upper limit on the size of the hematite core. (Note that the magnetization is dominated by the biggest particles.)

Additional support for that picture is given by the measurements of the volume dependence of the blocking temperature [79], Fig. 5.19. In this experiment particles were grouped according to their core volume by using chemical methods of changing the loading of iron into the ferritin cage. It is evident from Fig. 5.19 that T_B is proportional to V, in good agreement with the formula $T_B = E_a V / 30$.

The final confidence in our picture of ferritin comes from the AC measurements of the blocking effect. Below the Néel temperature the particles should appear magnetically ordered or superparamagnetic, depending on whether the period of the AC oscillations is small or large compared with the lifetime of a particular orientation of the magnetic moment. Consequently, the blocking temperature becomes dependent on the frequency of the AC field, ω, through $T_B = -U / \ln(\omega t_0)$. This is in apparent agreement with the experimental data [81] shown in Fig. 5.20. All data, including the Mössbauer ($\omega \simeq 4 \times 10^8$ Hz) and DC ZFC magnetization ($\omega \simeq 1$ Hz) measurements, fall nicely on a straight line.

As has been discussed in Chapter 3, antiferromagnetic materials are the best candidates for the study of quantum tunneling. They provide both tunneling

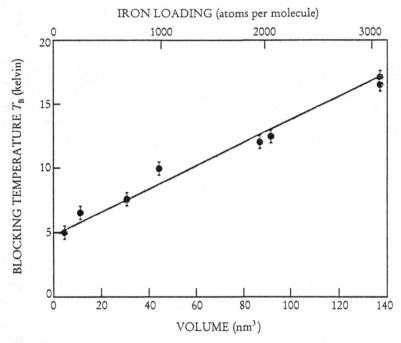

Figure 5.19 The dependence of the blocking temperature on the ratio of the volume of the core and the volume of the ferritin molecule. (Data from [79].)

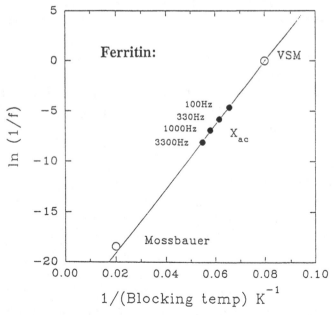

Figure 5.20 The shift of the blocking temperature with temperature in AC experiments. The points corresponding to the blocking observed in DC susceptibility and Mössbauer measurements are also included. (Data from [81].)

rates and crossover temperatures significantly higher than those of ferromagnets. Relaxation measurements on these materials, however, are more difficult than are those on ferromagnets because of their very low net magnetic moment. SQUID magnetometers, owing to their high sensitivity, made such measurements possible with only a few milligrams of material. Measurements of the magnetic relaxation in natural ferritin within four decades of time have demonstrated its purely logarithmic character [80], see Fig. 5.21. This is in accordance with the broad distribution of sizes deduced from the Mössbauer measurements [77,78]. Notice that only a small part of the total magnetization is relaxing below 4 K, suggesting that the relaxation is due to the smallest particles present. Bigger particles are blocked on the timescale of the experiment. The temperature dependence of the magnetic viscosity [80] is displayed in Fig. 5.22. The maximum in the viscosity at 8.5 K is consistent with the unblocking of the biggest particles and the crossover to the superparamagnetic regime whereby the $\ln(t)$ law is lost and the relaxation becomes very fast. Note that the maxima in the viscosity and in the ZFC magnetization occur at different temperatures because of the different timescales of these two experiments. The 8 nm particles, which are blocked at 13 K in the ZFC experiment (that lasts about 10 s), are blocked only at 8.5 K in the relaxation experiment (that lasts more than 1 h). In the range 2.3–6 K the viscosity changes linearly with temperature and it becomes

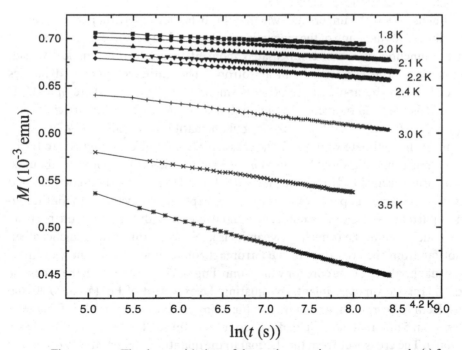

Figure 5.21 The decay with time of the total magnetic moment versus $\ln(t)$ for various temperatures obtained from ferritin molecules. (Data from [80].)

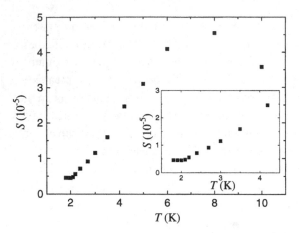

Figure 5.22   The magnetic viscosity extracted from the relaxation data as a function of temperature for the ferritin protein sample. (Data from [80].)

independent of temperature below 2.3 K. Extrapolation of the straight line for the range 2.3–6 K to lower temperatures would give zero viscosity at about 2 K. This would mean that all of the particles which contribute to the magnetization are blocked below 2 K. In its turn this would suggest that the sizes of these particles are bounded from below such that 2 K is the blocking temperature for the smallest size. For a 1 h relaxation experiment that would correspond to $d_{min} \simeq 4$ nm. In fact, however, this does not happen and the system continues to relax at a constant rate below 2.3 K.

Let us now try to understand whether this behavior of ferritin particles can be attributed to tunneling of the magnetic moment. As has been explained in Section 3.2, tunneling of a non-compensated moment in antiferromagnetic particles requires a non-zero transversal anisotropy. The antiferromagnetic dynamics occurs when the non-compensation is small, $m^2 \ll 2\chi_\perp K_\perp$, where $m = M/V$ and M is the total moment. Assuming that $K_\perp \simeq K_\parallel = 2.5 \times 10^5$ erg cm^{-3} and $\chi_\perp \simeq 2 \times 10^{-5}$ cm^{-3} G^{-1}, just like for bulk hematite, we obtain $(2\chi_\perp K_\perp)^{1/2} \simeq 3$ emu. If this estimate were valid, the crossover from antiferromagnetic to ferromagnetic tunneling should occur at $m \simeq 3$ emu, which corresponds to about 60 non-compensated Fe^{3+} ions for 8 nm particles and to about 15 non-compensated iron atoms for 5 nm particles. Given that the experimental value is 15 for the biggest particles and that this number should decrease with the size of the particle, we conclude that the tunneling dynamics is likely to be antiferromagnetic, albeit not far from the crossover to the ferromagnetic regime. We can now estimate the maximal size of the core for which tunneling can be exhibited on the timescale of the relaxation experiment. By equating the exponent of Eq. (3.66) to 30 and neglecting the m-term, we obtain that tunneling can occur in particles of diameter less than 5 nm, that is, for a loading into the ferritin shell of fewer than 1000 iron atoms. The crossover from the thermal to the quantum regime should occur at a temperature given by Eq. (3.67). On substituting into that formula $K_\parallel$ and $\chi_\perp$

values for bulk hematite, we obtain $T_c \simeq 5$ K, which is a rather high value to insure that the crossover can fall within the experimental range of temperatures. We, therefore, conclude that the viscosity data presented in Fig. 5.22 may indeed constitute evidence of spin tunneling in ferritin particles. We shall return to this question in Chapter 6 in connection with other tunneling experiments concerning ferritin.

Non-thermal magnetic relaxation has been reported to occur in 15 nm $Tb_{0.5}Ce_{0.5}Fe_2$ ferromagnetic particles [82]. In that work the relaxation measurements were performed down to 50 mK. The authors of [82] defined the average relaxation time as the time corresponding to a significant change in the total magnetization after the reversal of the field. They assumed that this time was proportional to $\exp(\langle U \rangle / T_{esc})$, where T_{esc} is the escape temperature introduced in Chapter 2 and $\langle U \rangle$ is some average barrier that depends on the applied field. Analysis of the data demonstrated that, above 4 K, T_{esc} was close to the physical temperature T. However, below 4 K the escape temperature was systematically higher than T. The plot of T_{esc} versus $\ln(T)$ is shown in Fig. 5.23.

The above examples illustrate some typical features of the magnetic relaxation in small particles. Many such systems have been studied. Kodama et al. [83] reported measurements on $NiFe_2O_4$ with an average particle size of 7 nm and dispersion of about 50%. At the highest field of 6 T the hysteresis loop remained open, indicating the presence of high anisotropy fields. A logarithmic time depen-

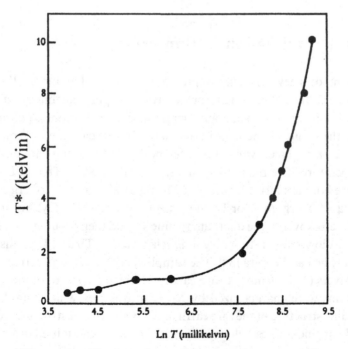

Figure 5.23 The dependence of the escape temperature, T_{esc}, on the natural logarithm of temperature for $Tb_{0.5}Ce_{0.5}Fe_2$ particles. (Data from [82].)

dence of the magnetization was observed from 1 min to several hours. The extracted viscosity exhibited a non-thermal behavior below 1 K. Ibrahim *et al.* [84] reported a non-thermal relaxation in 3 nm FeOOH particles. The ZFC magnetization data indicated that a significant fraction of the particles had anisotropy energies higher than 10^6 erg cm^{-3}. From the observed ln (t) relaxation the authors of [84] extracted the temperature dependence of the viscosity. Above 7 K the viscosity grew linearly with temperature, whereas below 7 K it was nearly independent of temperature. These and other similar observations were interpreted in terms of quantum tunneling.

Finally, we would like to comment that the dynamics of the magnetization in small particles may be more complicated than that described by our simple models. It is widely believed that, for particles whose size is small compared with the domain wall thickness, the metastable states decay via a uniform rotation of the magnetic moment. The argument is that any non-uniformity would cause a large exchange energy. The energy barrier for the uniform rotation is due to the magnetic anisotropy, $U = KV$. This seems to be in agreement with the experimental data on blocking and superparamagnetism in many systems. One should remember, however, that the spins on the surface of small particles have different exchange interactions than those inside the particle. If the surface were random, some of the outer spins could be loosely coupled to other spins. Different states of these surface spins may give rise to a number of small barriers inside the particle. In this case there will be no direct correspondence between the size of the particle and the height of the barrier.

5.4 Quantum relaxation in thin films

The theoretical models developed in Chapters 3 and 4 show that, among the ferromagnets, the most favorable system for the study of spin tunneling should be that which has a strong easy-plane anisotropy and two or more equilibrium orientations of the magnetic moment in the plane. In this case the dynamics of M occurs in the easy-plane, which enhances both the tunneling rate and the crossover temperature. The required symmetry is possessed by Tb and Dy. At low temperature both are ferromagnetic, $K_\perp$ being as high as 5×10^8 erg cm^{-3}, and $K_\parallel$ being 2.4×10^6 erg cm^{-3} for Tb and 7.5×10^6 erg cm^{-3} for Dy. According to Eq. (3.21) the crossover to the quantum regime should then occur in the kelvin range. A study of low-temperature magnetic relaxation in Tb and Dy films was conducted by O'Shea and Perera [85]. The sample, prepared by sputtering, consisted of thin layers (1–1000 nm) of rare earth separated by 18 nm of molybdenum. The systems were characterized by X-ray diffraction, which established the polycrystalline structure of the rare earth films. A significant structural disorder was found for thicknesses below 30 nm; at 10 nm no crystalline order was detected. The viscosity data are shown in Fig.5.24. One notices that the crossover to non-thermal relaxation for Tb occurs at about 20 K, which is the highest observed to date. This is in accordance with the theoretical expectation, although the value of the crossover temperature seems a little high. One may ask what tunneling inside rare earth films means. The answer to this question is not known at the moment. If the viscosity data on Tb and Dy films indeed reflect processes of quantum tunneling of the magnetization, one may think about quantum nucleation of magnetic domains, depinning of domain walls, or rotation of the magnetic moment in small clusters. For that reason it would be important to establish the mechanism of the magnetic relaxation in these materials at high temperature. This question also concerns the data presented below.

A very clear crossover from thermal relaxation to temperature-independent relaxation has been observed in $TbFe_3$, $SmFe_4$, and $SmCo_4$ thin films [86,87].

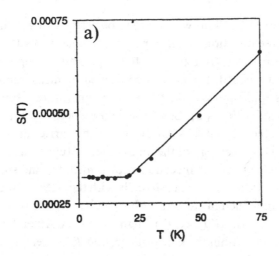

Figure 5.24 The temperature dependence of the magnetic viscosity for (a) Tb thin films of 77 nm thickness, and (b) Dy thin films of 90 nm thickness. (Data from [85].)

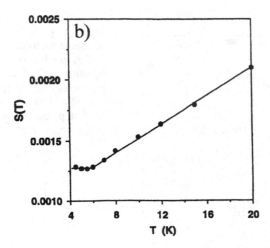

The films were prepared by electron-beam evaporation onto kapton foil substrate. The procedure was described in [87]. The terbium–iron film, for example, was prepared by evaporating 10 nm of silver buffer onto kapton, followed by 0.3 nm of iron and 0.4 nm of terbium, then repeating the Fe–Tb evaporation 50 times to obtain a detectable magnetic signal. The high-angle X-ray diffraction data revealed only peaks corresponding to the crystalline silver; the magnetic layer, therefore, was presumed amorphous. This was confirmed by Mössbauer studies in the 4–300 K range. The final composition of the magnetic layer, $TbFe_3$, was obtained by scanning electron-probe analysis. FC and ZFC magnetization data, Fig. 5.25, indicate the presence of widely distributed metastable magnetic states with blocking at around 300 K. It is most likely that they are related to the orientation of ferromagnetic clusters with respect to the applied field and effective anisotropy axes. From the hysteresis curve anisotropy fields of more than 5 T were detected, which must put quantum effects within the kelvin range for this material.

The magnetic relaxation in $TbFe_3$ films was found to be perfectly logarithmic from a few seconds to 2 h (the duration of the experiment). The viscosity data are shown in Fig. 5.26. The crossover from thermal to temperature-independent relaxation at 6 K is very well defined. It has been confirmed by measurements on many samples at various applied fields. The crossover temperature decreases with the field, as is illustrated by Fig. 5.27. Since the field reduces the barriers, this is in accordance with the theory that says that the lower the barrier the smaller T_c. This latter statement is independent of the mechanism of tunneling.

Even without exact knowledge of the mechanism of tunneling, one should notice that experimental observations correlate strongly with theoretical models. A small part of the total M (confined in clusters with small barriers) is relaxing, the relaxation follows the $\ln(t)$ law, $S(T) \propto T$ at high T and is independent of T at low T, the crossover occurs within the kelvin range, and T_c decreases when

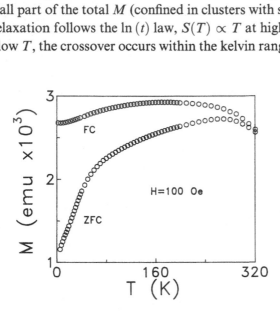

Figure 5.25 Low-field magnetization curves for $TbFe_3$ thin film obtained by using the ZFC and FC processes with the applied field $H = 0.01$ T parallel to the film plane. (Data from [87].)

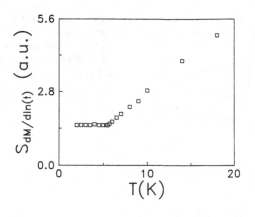

Figure 5.26 The low-temperature dependence of the magnetic viscosity for $TbFe_3$ thin film. (Data from [87].)

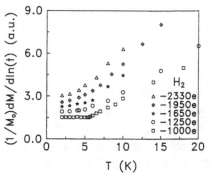

Figure 5.27 The variation of the magnetic viscosity with the value of the magnetic field. The sample was cooled in the field $H_1 = 0.01$ T and then the field was changed to a new value, H_2. (Data from [87].)

barriers are lowered by the field. These points are clearly in favor of the tunneling interpretation.

Arnaudas *et al.* [88] reported a similar behavior in Tb_2Fe amorphous films. This material has an extremely high magnetic anisotropy, $H_a = 9.5$ T, which is likely to dominate its magnetic structure. The magnetization inside the film must point along the local anisotropy axis. In this case, the crossover from the thermal to the quantum regime is determined by Eq. (3.67) [89]. For experimental values of $K = 9 \times 10^7$ erg cm^{-3} and $\chi = 8 \times 10^{-3}$ it gives $T_c \simeq 7$ K. The magnetic viscosity is shown in Fig. 5.28. It can be seen from Fig. 5.28 that the crossover occurs at about 8 K. From the value of the tunneling rate, the authors of [88] estimated that transitions occur in volumes of size 3 nm^3 that contain about 100 spins.

5.5 Quantum relaxation in bulk magnets

Non-thermal magnetic relaxation has also been observed in measurements of bulk magnetic crystals. If this were due to quantum tunneling of the magnetization, the most natural assumption would be that quantum depinning of domain

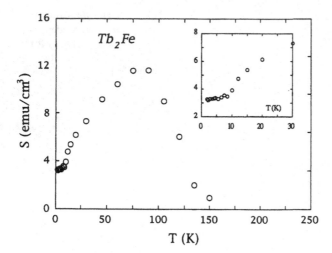

Figure 5.28 The low-temperature magnetic viscosity versus temperature in Tb_2Fe film. (Cited from [88].)

walls were involved. One of the materials that has been studied intensively is single-crystalline $SmCo_{3.5}Cu_{1.5}$. This alloy is characterized by a significant chemical disorder that produces inhomogeneities of size 5–50 μm. The chemical disorder is believed to be responsible for strong pinning of domain walls. $SmCo_{3.5}Cu_{1.5}$ has a very high uniaxial anisotropy, $H_a \simeq 5$ T. Assuming that disordered regions create perpendicular anisotropy of the same order, one would expect (see Section 3.4) that, below a few kelvins, the depinning of domain walls should be dominated by quantum tunneling. An anomaly in the magnetic relaxation of this material at low temperature, below 50 K, was first reported by Barbara and Uehara in 1977 [90], and later studied in more detail [91]. The data were analyzed in terms of the average relaxation time for barriers of different height. The authors of [91] then assumed that this average time depended exponentially on the escape temperature T^* (the same as our T_{esc} of Section 3.1). T^* must coincide with T in the thermal regime but is a non-zero constant in the limit of $T \to 0$. This is a rather crude approximation that may, nevertheless, describe the anomaly in the low-temperature relaxation. The dependence of T^* on T for $SmCo_{3.5}Cu_{1.5}$ is shown in Fig. 5.29. Another interesting observation made in this material is a staircase hysteresis loop [92]. Giant steps in the hysteresis loop emerge when the temperature is lowered below 2 K, their location being reproducible when cycling the magnetic field. The study of the polar Kerr effect has shown that each step is associated with the depinning of a domain wall, which takes place in some microscopic region of the sample and then results in the avalanche growth of the domain. This observation has been interpreted [92] as an indication of tunneling. The steps, however, could also be due to a series of critical fields needed to depin domain walls at a few 'weak points' (where the anisotropy is lowered by disorder).

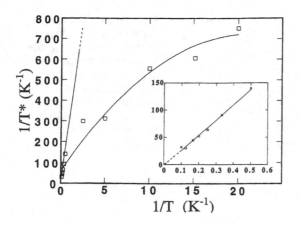

Figure 5.29 The low-temperature anomaly in the magnetic relaxation of $SmCo_{3.5}Cu_{1.5}$. (Cited from [91].)

Another bulk material that exhibits a non-thermal relaxation is $TbFeO_3$. Single crystals of this optically transparent orthoferrite have been studied for years. Its magnetic structure is well known from magnetic measurements [93,94], neutron diffraction [95], and Mössbauer data [96]. Depending on the temperature interval, it has different antiferromagnetic orderings with a small canting of equivalent sublattices that results in weak ferromagnetism. The Néel temperature is 680 K. Down to 10 K iron is ordered antiferromagnetically whereas terbium is paramagnetic. Below 10 K both Fe and Tb are ordered in antiferromagnetic structures, iron sublattices being slightly canted and terbium sublattices collinear. The magnetic relaxation in terbium orthoferrite is due to the motion of antiferromagnetic domain walls.

Figure 5.30 shows the time dependence of the low-temperature magnetic relaxation in a single crystal of $TbFeO_3$ [97]. The new feature is the nearly exponential time decay of the magnetization, in contrast to the logarithmic relaxation widely observed in other systems. As has been discussed in Section 4.2, this is an indication of a very narrow distribution of barrier sizes. Note that even a 2% dispersion in the barrier height would completely destroy the exponential law. One can talk, therefore, about a single barrier for the depinning of domain walls at low temperature. Writing the escape rate in the form

$$\ln\left(\Gamma\right) = \ln\left(\nu\right) - \frac{U(H)}{T_{\text{esc}}(T)}, \qquad (5.20)$$

one can plot T_{esc} versus T, as has been done in Fig. 5.31. Above 2.3 K, T_{esc} coincides with the experimental temperature, implying that the depinning of domain walls is purely thermal. The barrier calculated from that regime decreases with increasing field and is about 60 K at 100 Oe. Below 2.3 K a dramatic departure from thermal relaxation takes place, with T_{esc} becoming nearly independent of temperature.

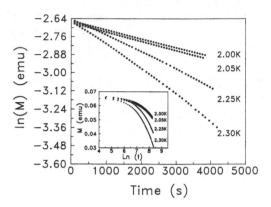

Figure 5.30 The time dependence of the logarithm of the magnetization at various temperatures after the field in the *c*-direction had been switched from 20 Oe to −200 Oe. The inset shows *M* versus ln (*t*). (Cited from [97].)

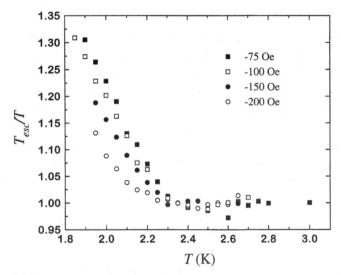

Figure 5.31 T_{esc}/T versus T for a single crystal of TbFeO$_3$.

Because the terbium orthoferrite apparently possesses a single barrier for the depinning of domain walls, it provides an interesting example for the study of macroscopic quantum tunneling. The question of the nature of the barrier of course remains. A simple estimate shows that individual atomic impurities cannot provide the required barrier height. On the other hand, clusters of impurities would have a distribution of sizes and, thus, could not provide a single barrier. The most plausible explanation could be the pinning of domain walls by twin boundaries. The model of Section 3.4 should be a good description of this case, provided, of course, that antiferromagnetic dynamics is involved. As has been discussed in Section 3.4, the crossover from the thermal to the quantum regime for the depinning of anti-ferromagnetic domain walls is governed by the same formula, Eq. (3.67), as that for tunneling in antiferromagnetic particles. On substituting into this formula the experimental values $K = 10^5$ erg cm^{-3} for the anisotropy constant and $\chi = 10^{-4}$ for the magnetic susceptibility, one obtains $T_c \simeq 2$ K which is in agreement with experiment.

Our last example is single-crystalline BaFe$_{10.2}$Sn$_{0.74}$Co$_{0.66}$O$_{19}$ [98]. The hexagonal BaFe$_{12}$O$_{19}$ is a ferrimagnet with uniaxial anisotropy along the c-axis. In pure non-substituted large single crystals of this hexaferrite, domain walls move freely in response to changes in the magnetic field. In order to study long-time relaxation it is necessary, therefore, to introduce some substitution of Co and Sn by Fe to produce pinning centers. Figure 5.32 shows the hysteresis loop for pure single-crystalline BaFe$_{12}$O$_{19}$ and for single-crystalline BaFe$_{10.2}$Sn$_{0.74}$Co$_{0.66}$O$_{19}$. In the case of the pure hexaferrite, both the coercive field and the remanent magnetization are zero due to the absence of barriers. Introducing Co and Sn into the ferrimagnetic structure pins the domain walls and gives rise to magnetic relaxation, that is, both to coercivity and to remanence. The sample was field cooled, in $H = 10$ Oe, applied in the c-axis direction, from 300 K down to the measuring temperature and then the field was changed to a new value, $H = -220$ Oe. The magnetic moment follows the time-logarithmic law very well and the viscosity values deduced from the data are shown in Fig. 5.33. At temperatures higher than 3 K the viscosity behaves linearly with respect to T but as the temperature goes to zero the viscosity remains nearly constant. The theoretical prediction

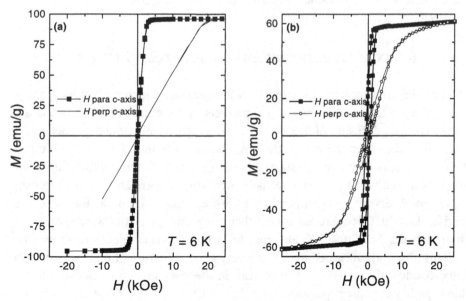

Figure 5.32   The hysteresis loop at $T = 6$ K with the field applied perpendicular to the c-axis. (a) single-crystalline BaFe$_{12}$O$_{19}$ and (b) single-crystalline BaFe$_{10.2}$Sn$_{0.74}$Co$_{0.66}$O$_{19}$. (Cited from [98].)

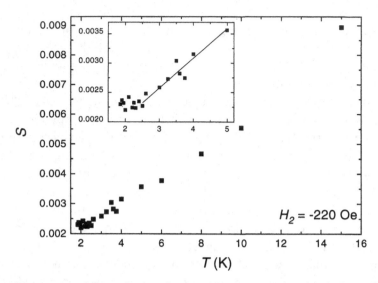

Figure 5.33   Magnetic viscosity data versus temperature for the low-temperature regime. (Cited from [98].)

for the crossover from the thermal to the quantum regime for the domain-wall motion in antiferromagnets is given by Eq. (3.67). On substituting into this equation the values $K = 2.6 \times 10^6$ erg cm−3 and $\chi_\perp = 2 \times 10^{-7}$ measured for the substituted hexaferrite we obtain $T_c \simeq 1.5$ K, in qualitative agreement with the obtained experimental value of 2.5 K.

5.6   Relaxation experiments in superconductors

A very similar situation occurs in high-temperature superconductors [57, 99–109]. The Anderson–Kim model [61] predicts that the magnetic relaxation due to the thermal diffusion of flux lines follows a logarithmic time dependence, the viscosity being proportional to the absolute temperature. This is a well-established feature of conventional superconductors. For high-temperature superconductors a significant anomaly in the temperature dependence of the viscosity has been observed. Low-temperature measurements on granular Ba–La–Cu–O and Sr–La–Cu–O [100] revealed that their magnetic viscosities were nearly constant between 5 mK and 1 K, and grew linearly with the temperature above 1 K.

This observation turned out to be a very general feature of high-temperature superconducting materials. The plateau in the viscosity was observed for the high-quality ceramic superconductors Y–Ba–Cu–O, Tl–Ba–Ca–Cu–O, and Bi–Sr–Ca–Cu–O [101–103], as well as for organic [104], heavy-fermion [105], and Chevrel-phase superconductors [106]. Figure 5.34 shows the temperature depen-

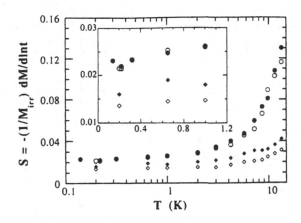

Figure 5.34 The temperature dependence of the magnetic viscosity of $Bi_2Sr_2CaCu_2O_x$. (Data from [100].)

dence of the magnetic relaxation in a single crystal of $Bi_2Sr_2CaCu_2O_x$ [101]. The plateau in the magnetic viscosity below 1 K is quite appealing. The Bi-based material is one of the most anisotropic high-temperature superconductors. To estimate the probability of quantum depinning of a pancake vortex one can use the expression derived in Chapter 4. Taking 15 nm for the distance between CuO_2 planes and 30 $\mu\Omega$ cm for the normal state's resistivity, one obtains a reasonable number, $B = 30$, for the WKB exponent. The pinning energy found from high-temperature measurements was about 240 K for this sample. Then, $T_c \simeq U/B$ with B of Eq. (4.77) gives a few kelvins for the crossover from the thermal to the quantum regime, in agreement with experiment.

Tunneling of pancake vortices should not be surprising because these are very small (of the size of a few lattice spacings) and light (of mass comparable to the electron mass) objects. This situation is radically different from that in bulk magnetic materials because there the tunneling entities, domain walls, are always extended objects. Of course, the picture of pancake vortices rather than flux lines arises only from the high anisotropy of copper oxides. For that reason, the more anisotropic the superconductor the more pronounced the quantum effects that should be expected. This is reflected in the data on the magnetic viscosity of Tl-2212, the most anisotropic high-temperature superconductor. For this material the ratio of penetration lengths in two perpendicular directions is as big as 350, compared with 200 for the Bi-based material discussed above. The low-temperature magnetic viscosity of this material is shown in Fig. 5.35. The crossover to the quantum regime occurs at about 8 K. Similarly high crossover temperatures have been observed in other Tl-containing superconductors. Insofar as tunneling is concerned, the dissipation apparently plays a more important role in superconductors than it does in magnets. Whereas in magnets the overdamped regime is unlikely, in high-temperature superconductors it is natural.

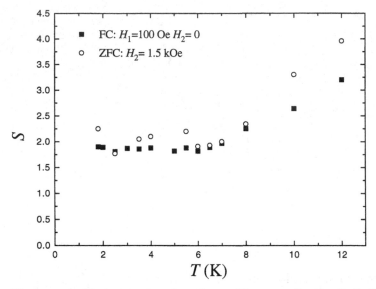

Figure 5.35   The temperature dependence of the magnetic viscosity for Tl-2212 as a function of temperature corresponding to (a) the ZFC process with the applied field $H = 1.5$ kOe and (b) remanent magnetization. (Data from [108].)

5.7   Additional remarks on relaxation experiments

The systems which exhibit the plateau are very different: ensembles of small particles, bulk crystals, thin films, random magnets, etc. The theory suggests that the crossover from the thermal to the quantum regime occurs at a temperature that scales linearly with the anisotropy field and the order of magnitude of which is given by $\mu_B H_a$. This correlates with the experimental data, Fig. 5.36 and Table 5.1, on all materials studied to date. It is hard to find any correlations, other than tunneling, among the materials presented in Fig. 5.36, which would explain the data in terms of thermal transitions.

For future experiments it is useful to mention effects that could be misinterpreted as quantum relaxation. Two different populations of barriers can mimic the plateau in the viscosity when measurements have not been performed down to a low enough temperature. This can be illustrated by considering a system consisting of two species: $CoFe_2O_4$ particles mixed with γ-Fe_2O_3 particles, both of mean diameter 3.5 nm, in a potassium silicate matrix [110]. The FC and ZFC magnetization curves of this system are shown in Fig. 5.37. That there are two populations of particles is clearly seen from the existence of two maxima in the ZFC curve. These maxima correspond to the independent blocking of the two species and are roughly in accordance with the difference in their anisotropy energies; the lower maximum corresponds to the blocking of γ-Fe_2O_3 particles. The

Table 5.1   List of materials considered in Fig. 5.36

Number	Circles	Squares	Triangles
1	FeOOH	$(Tb_{0.2}GD_{0.8})_2Cu$	Tb
2	$NiFe_2O_4$	TB_2Cu	Dy
3	Ferritin	Tb_2Fe	
4	$\gamma\text{-}Fe_2O_3$	Fe_3Tb	
5	$BaFe_{12}O_{19}$	Co_4Sm	
6	Fe_3O_4	Fe_4Sm	
7	CrO_2	CoCu	
8	FeC	DyCu	
9	$CoFe_2O_4$	$Tb_{0.25}Y_{0.75}Al_2$	

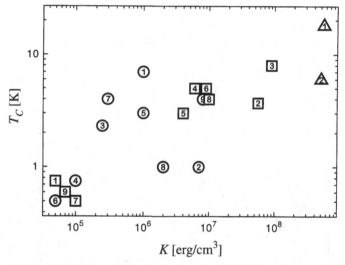

Figure 5.36   The crossover temperature versus the anisotropy field of the materials listed in Table 5.1.

viscosity data for this system obtained after the field had been switched from 4 kOe to −4 kOe are displayed in Fig. 5.38. The maximum in the viscosity at 2.5 K corresponds to the blocking of $\gamma\text{-}Fe_2O_3$ particles when the barriers are reduced by the 4 kOe field, and is in accordance with the ZFC data. Figure 5.38 shows that, if the viscosity data were taken down to only 3 K, without consulting the ZFC data, the low-temperature behavior could easily be confused with the

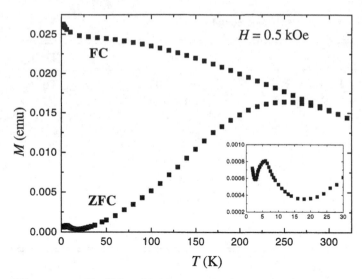

Figure 5.37 The FC and ZFC magnetization curves of the mixture of $CoFe_2O_4$ and γ-Fe_2O_3 particles. The inset emphasizes the low-temperature behavior. (Cited from [110].)

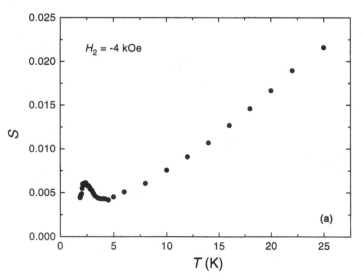

Figure 5.38 The low-temperature magnetic viscosity of the mixture of $CoFe_2O_4$ and γ-Fe_2O_3 particles obtained after the field had been changed from 4 kOe to −4 kOe. (Cited from [110].)

plateau. In fact the plateau does appear for the above system in high fields, when the barriers in γ-Fe_2O_3 particles are destroyed by the field, Fig. 5.39. The field of 8 kOe is higher than the anisotropy field of γ-Fe_2O_3 particles and thus eliminates their contribution to the slow relaxation. This is in accordance with the

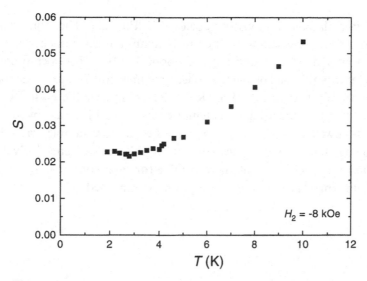

Figure 5.39 The low-temperature magnetic viscosity of the mixture of $CoFe_2O_4$ and γ-Fe_2O_3 particles obtained after the field had been switched from 4 kOe to −8 kOe. (Cited from [110].)

fact that the ZFC data taken in the 8 kOe field exhibit no low-temperature maximum. The plateau in Fig. 5.37 can be attributed to tunneling in $CoFe_2O_4$ particles in accordance with the data obtained for these particles alone, see Fig. 5.16. The lesson to be learned from this example is that the viscosity data should be taken together with the ZFC data down to the lowest temperature available and for various values of the applied field.

Self-heating of a sample due to the decay of metastable states can erroneously be taken for a non-thermal relaxation [111]. Owing to this effect the temperature of the interior of the sample may be higher than the temperature of the sample surface, where the temperature is measured. This higher temperature can be confused in an experiment with the escape temperature introduced above. The self-heating can easily be calculated. Indeed, the rate of the energy input into the sample when the system is relaxing is $H \cdot (dM/dt)$, where H includes the external and the demagnetizing field. This should be introduced as a source into the equation for the heat conductivity describing the thermal balance inside the sample. The difference between the temperature of the interior and the surface then depends on how fast the heat is taken away owing to the thermal conductivities of the sample and the bath. The self-heating calculated for the experiments discussed in this chapter turns out to be negligible in the kelvin range but can become of concern in the millikelvin range because of the rapid fall of the thermal conductivity with decreasing temperature. An estimate of the upper bound on the self-heating can be made without solving the equation of the heat conductivity. To illustrate this, let us stick to the example of the relaxation in Fe_3Tb, which exhibits a plateau in the viscosity below 6 K, Fig. 5.26. The total change in the magnetization of the sample during the relaxation measurement was about 10^{-4} emu. For a demagnetizing field of about 1000 Oe this corresponds to 0.1 erg total energy released owing to the decay of metastable states. The specific heat of the sample was about 15 erg K^{-1}. Assuming that the entire released energy remains inside the sample, its temperature should increase by less than 10 mK. This, however, is a huge overestimate because the sample was a 30 nm thin metallic film in a helium vapor bath that should cool very effectively during the relaxation process. For a bulk non-metallic (or superconducting) sample in the millikelvin range, a more careful analysis may be needed.

Chapter 6

Other experimental approaches

6.1   AC susceptibility and noise measurements

In recent years some AC susceptibility measurements the purpose of which was to nail the effect of magnetic tunneling have been reported [112–115]. Before describing these experiments we shall briefly discuss the theory behind them. Basically (except for the resonance experiment discussed later) they were variants of relaxation measurements. Let us assume that we have a system of identical non-interacting particles (or clusters) whose total magnetization is relaxing. In the absence of a magnetic field, $\tau(\mathrm{d}M/\mathrm{d}t) + M = 0$ is satisfied, where τ is the single relaxation time. When, on the other hand, no relaxation is present but the system is subject to a slowly oscillating weak magnetic field, $h = h_0 \cos(\omega t)$, then $M(t) = \chi_0 h_0 \cos(\omega t)$. In the presence of both relaxation and an AC field,

$$\tau \frac{\mathrm{d}M}{\mathrm{d}t} + M = \chi_0 h_0 \cos(\omega t). \tag{6.1}$$

By writing the solution in the form

$$M = \chi' h_0 \cos(\omega t) + \chi'' h_0 \sin(\omega t), \tag{6.2}$$

we obtain

$$\chi' = \chi_0 \frac{1}{1 + (\omega\tau)^2}$$
$$\chi'' = \chi_0 \frac{\omega\tau}{1 + (\omega\tau)^2}. \tag{6.3}$$

for the in-phase and out-of-phase susceptibilities correspondingly.

The relaxation time τ depends on the volume of the particle, $\nu\tau = \exp\left(KV/T\right)$. Turning now to the distribution of volumes, we should integrate Eq. (6.3) taking account of the fact that the contribution of each particle to the total moment, and, thus, the total susceptibility, is proportional to its volume. This gives

$$\chi' = \frac{\chi_0}{V_{\text{tot}}} \int_0^\infty dV \frac{f(V)V}{1 + [\omega\tau(V)]^2}$$

$$\chi'' = \frac{\chi_0}{V_{\text{tot}}} \int_0^\infty dV \frac{f(V)V\omega\tau(V)}{1 + [\omega\tau(V)]^2}, \tag{6.4}$$

where $V_{\text{tot}} = \int_0^\infty dV f(V)V$ is the total magnetic volume and all susceptibilities are now referred to the unit volume of the magnetic material. Similarly to our analysis of DC relaxation measurements, let us introduce $V_{\text{B}} = (T/K)\ln\left(\nu/\omega\right)$. At low temperature, the factor $\{1 + [\omega\tau(V)]^2\}^{-1}$ in the definition of χ' is equivalent to the theta-function of $V_{\text{B}} - V$, which gives

$$\chi' = \chi_0 \frac{\int_0^{V_{\text{B}}} dV f(V)V}{\int_0^\infty dV f(V)V}. \tag{6.5}$$

Insofar as the factor $\omega\tau(V)/\{1 + [\omega\tau(V)]^2\}$ in the definition of χ'' is concerned, it rapidly goes to zero as soon as the value of V departs from V_{B}. This allows one to approximate χ'' by

$$\chi'' = \frac{\chi_0 V_{\text{B}} f(V_{\text{B}}) \int_0^\infty dV \, \omega\tau(V)\{1 + [\omega\tau(V)]^2\}^{-1}}{\int_0^\infty dV f(V)V}. \tag{6.6}$$

By using the explicit form of $\tau(V)$, the integration over V can be performed explicitly. This gives

$$\chi'' = \frac{\pi}{2}\chi_0 \frac{T}{K} \frac{V_{\text{B}} f(V_{\text{B}})}{\int_0^\infty dV f(V)V}. \tag{6.7}$$

It is also useful to notice the relation between χ' and χ'' that follows from the above formulas,

$$\chi'' = -\frac{\pi}{2}\frac{\partial\chi'}{\partial\ln\omega}. \tag{6.8}$$

These formulas suggest a few experimental tests. First, χ'/χ_0 should depend on the temperature and frequency via the combination $T\ln\left(\nu/\omega\right)$. This statement is analogous to the statement of the $T\ln\left(\nu t\right)$ dependence of the moment in the DC relaxation measurement. Consequently, the presence of quantum tunneling should reveal itself in the departure from the above scaling at low temperature. In the presence of quantum tunneling, the temperature in the definition of V_{B} must be replaced by $T_{\text{esc}}(T)$. Secondly, the dependence of χ'' on temperature must show a clear crossover from the thermal to the quantum regime. For, e.g., the case in which $f(0) \neq 0$, the out-of-phase susceptibility must be proportional to the temperature in the thermal regime and inversely proportional to the temperature in the quantum regime owing to the superparamagnetic behavior of χ_0 (that is, $\chi_0 \propto 1/T$ for small applied magnetic fields). A comparison of Eq. (5.14) with Eq. (6.7) shows that, in the case in which the magnetic viscosity does not depend on the time (that is, the relaxation is linear in $\ln{(t)}$), the out-of-phase component of the susceptibility should be a frequency-independent one. This statement may not be easy to test because of the slow logarithmic dependence of χ'' on the frequency, however. If the period of the AC field has the order of magnitude of the DC relaxation time, the out-of-phase susceptibility and the viscosity are connected by the simple relation

$$\chi'' = \frac{\pi}{2}\chi_0 S. \tag{6.9}$$

Of course, the reader should remember that our models of AC susceptibility and viscosity only apply to the case of non-interacting particles. To be precise, the interaction should be small compared with the energy barriers involved.

When measurements are performed on a sample obtained by cooling in a zero DC field, and no DC field is applied thereafter, the sample must be close to thermal equilibrium. In this case χ'' is related to the noise spectrum $R(\omega)$ via the fluctuation–dissipation theorem, $R(\omega) = (2T/\omega)\chi''(\omega)$. This provides an independent test of how close the system is to thermal equilibrium.

The relation between χ'' and the magnetic viscosity, S, has been tested in two different materials. One is the ferritine sample, discussed in Chapter 5, which consists of weakly interacting magnetic particles and the other is a $Cu_{90}Co_{10}$ compound [115] formed by small magnetic clusters.

Experimental curves of $\chi'(T)$ and $\chi''(T)$ for the ferritin sample in the range 1.8–30 K measured at 9 Hz are shown in Fig. 6.1. The maximum observed in the two susceptibilities, in and out of phase, corresponds to the blocking of the magnetic moments. The extracted value of T_B agrees with the data both for the zero-field-cooled magnetization and for the magnetic viscosity, shown in Figs. 5.18 and 5.22, respectively. From the susceptibility data obtained at various frequencies, the validity of the mathematical relation between χ' and χ'' can be verified, see Fig. 6.2 [116]. It can also be verified that, at temperatures higher than 2 K, the data for χ'' scale as $T\log(\nu/\omega)$, see Fig. 6.3. The data on the temperature dependence of the magnetic viscosity, Fig. 5.22, and the out-of-phase susceptibility, Fig. 6.3, are in agreement with Eq. (6.9). Thus, the AC susceptibility measurements of ferritin support the hypothesis that, below 2 K, the magnetic relaxation in this material is due to quantum tunneling.

The example of $Cu_{90}Co_{10}$ is the one in which tunneling was not seen [115]. The bulk sample was obtained by planar flow casting. X-ray diffractometry showed that Cu and Co did not mix. Instead Co formed FCC clusters of average size 4 nm. The blocking temperature of the compound was 16 K. The magnetic

viscosity in the 1.8–20 K temperature range is shown in Fig. 6.4. This behavior of
the viscosity is classical, reflecting the behavior of the ZFC magnetization
curve. It corresponds to the progressive blocking of smaller and smaller particles
as the temperature decreases. This should be compared with the temperature

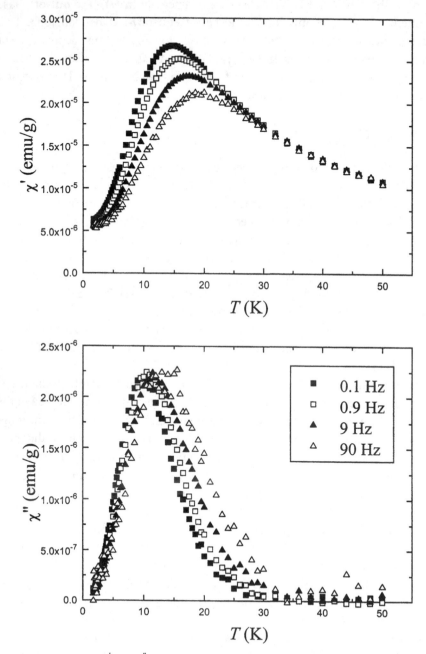

Figure 6.1 χ' and χ'' versus T at various frequencies for the ferritin sample. (Data from [116].)

dependence of χ'' shown in Fig. 6.5. This behavior is also classical; no crossover to the quantum regime discussed above has been detected. It has also been independently checked by measurement of the magnetic noise and good agreement between the two measurements has been obtained. In Fig. 6.6 we show the

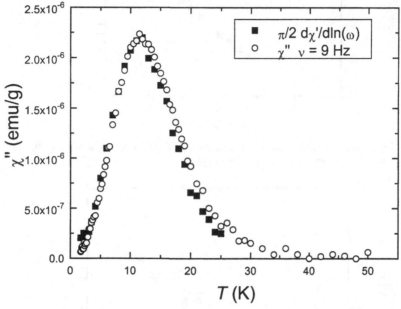

Figure 6.2 χ'' and $\partial\chi'/\partial\ln(\omega)$ versus T for the ferritin sample (Data from [116]).

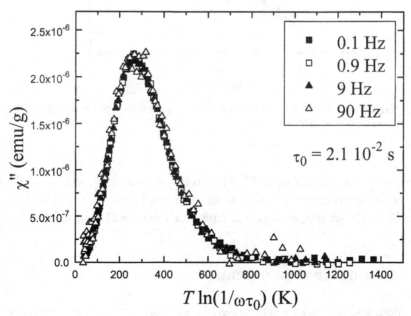

Figure 6.3 χ'' versus $T\ln(\nu/\omega)$ for the ferritin sample. (Data from [116].)

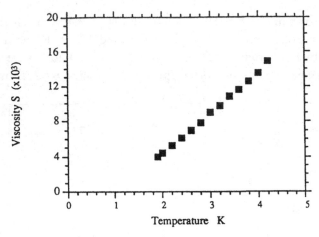

Figure 6.4 The temperature dependence of the magnetic viscosity of the $Cu_{90}Co_{10}$ compound. (Cited from [115].)

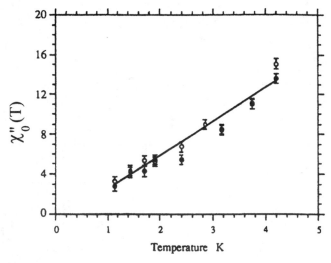

Figure 6.5 The low-temperature behavior of the out-of-phase susceptibility of $Cu_{90}Co_{10}$. (Cited from [115].)

temperature dependences of $S(T)/T$, directly measured χ'', and χ'' deduced from the noise-spectrum measurements. The behaviors of the three curves are qualitatively similar, as expected from our theoretical arguments.

6.2   Resonance experiments

Awschalom and co-workers made a significant effort [13,79] to observe MQC of the magnetic moment in ferritin. The structure of ferritin nanomagnets has

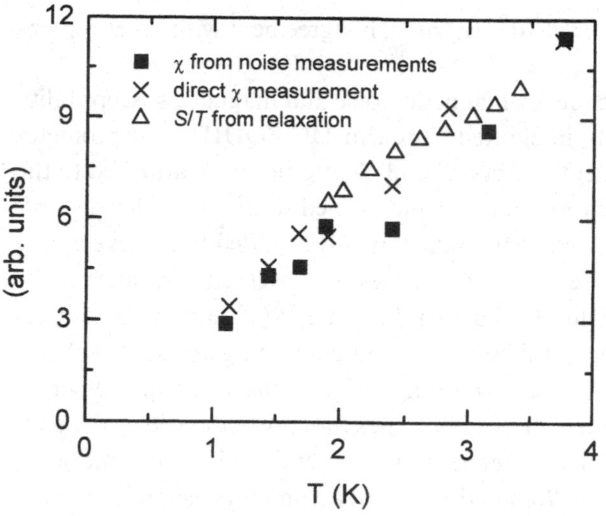
χ from noise measurements
direct χ measurement
S/T from relaxation
(arb. units)
T (K)
0
3
6
9
12
0
1
2
3
4

Figure 6.6 The temperature dependences of $S(T)/T$ and of the out-of-phase susceptibilities of $Cu_{90}Co_{10}$. (Cited from [115].)

already been discussed in Chapter 5, in connection with magnetic relaxation measurements [80]. These measurements, performed after Awschalom's experiment, supported the possibility of magnetic tunneling in ferritin. The observation of MQC is 'exponentially' more demanding though. Various samples were synthesized with the intention of obtaining ferritin particles of the same magnetic volume for each sample. The frequency dependence of the AC susceptibility was measured with the help of a sophisticated experimental technique. The goal was to observe a resonance in the absorption spectrum of ferritin that corresponds to the tunneling splitting of the ground state level, Fig. 3.13. A resonance was in fact observed and had a qualitatively correct dependence on the temperature, field, dilution, and size of the particles. This work received much attention in the literature and some important criticisms were raised. We will discuss here all facts that are for and against the tunneling interpretation of the observed resonance.

Natural ferritin has been used and a series of artificial ferritin samples has been synthesized, beginning with the empty apoferritin shell. The artificial samples used in measurements contained the same antiferromagnetic material as that which constitutes the natural ferritin core, with average loadings into the apoferritin shell of 100, 250, 500, 1000, 2000, 3000, and 4000 iron ions. The samples were characterized by chemical and biological analyses and transmission electron microscopy. The width of the distribution of diameters for each sample was in the range 10–25% [79]. ZFC magnetization curves were very similar to that shown in Fig. 5.18, with the blocking temperature in the range 5–15 K, depending on the iron loading. The anisotropy constant extracted from these measurements was of the order of 10^5 erg cm^{-3}, in agreement with other reports [76,80].

Measurements of high-frequency magnetic noise and magnetic susceptibility were performed using a fully integrated thin-film DC SQUID susceptometer [13]. The single-chip experiment and associated electronics were attached to the mixing chamber of a dilution refrigerator and cooled to 20 mK. The susceptometer assembly was electronically and magnetically shielded to achieve a very low level of magnetic flux noise. Figure 6.7 shows the observed resonance at the frequency $\nu_{\mathrm{res}} = \omega/(2\pi) \simeq 940$ kHz both in the noise, $R(\omega)$, and in the out-of-phase susceptibility for a 1000 : 1 diluted solution containing about 38 000 natural ferritin particles at 29.7 mK. According to [13], the resonance emerges from the background at about 0.2 K and becomes more pronounced as the temperature is lowered. The resonance frequency was essentially independent of the temperature. Measurements of $R(\omega)$ and $\chi''(\omega)$ on an undiluted sample failed to show any resonant structure. The dependence of the resonance frequency on the magnetic volume (iron loading) of ferritin particles is shown in Fig. 6.8. The volume V_0 corresponds to 4500 spins per protein molecule, which was the average volume of natural horse-spleen ferritin used in that study. The theory of tunneling in antiferromagnetic particles (see Chapter 4) predicts an exponential dependence of the tunneling rate on the volume. The data shown in Fig. 6.8

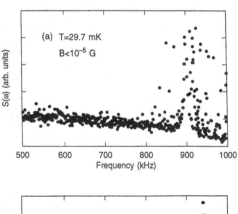

Figure 6.7 (a) The magnetic noise spectrum of a diluted sample of ferritin, $T = 19.7$ mK, $B = 5$–10 G. (b) The frequency-dependent magnetic susceptibility of the same sample. (Cited from [13].)

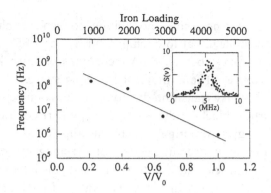

Figure 6.8 The dependence of the resonance frequency, ν_{res}, on the particle size, V/V_0. (Data from [79].)

seem to be consistent with that prediction. This could be strong evidence in favor of the tunneling interpretation. Indeed, it is rather difficult to imagine an oscillating mode, other than MQT, the frequency of which could drop exponentially with the total number of magnetic atoms. The fact that the resonance emerges only below some temperature is also in favor of the tunneling interpretation since one has to go below the crossover from the thermal to the quantum regime. Further support for the MQC interpretation comes from the facts that the observation of the effect requires a very low external field, $H \leq 1$ nT, and a well-diluted sample. This is needed to provide the condition of the degeneracy of the ground state.

Let us now discuss the obstacles to the MQC interpretation of the observed resonance in ferritin. Anupam Garg [72] pointed out that the power absorption in the case of MQT must be significantly lower than that which is detected [13], of order 10^{-24} W instead of the observed 10^{-21} W. Although this is a serious objection, we do not consider it crucial, given the theoretical, experimental, and electronics uncertainties involved. Anupam Garg also noticed that the decoherence arising from, e.g., nuclear spins of Fe ions must decrease the effect even further. One can try to rebuff this objection by citing the recent surprising observation [117] that coherent quantum oscillations of large spin are not destroyed by the interaction with a finite number of degrees of freedom. There exist, however, a few more trivial discrepancies between theory and experiment, which one may wish to resolve before addressing the subtle question of decoherence.

The fact that most of the particles become blocked below a few kelvins (Fig. 5.18), seems to be in disagreement with the statement that they tunnel with rates ranging from megahertz to gigahertz in the millikelvin regime. If tunneling were independent of temperature, it would have to reveal itself in the kelvin range as much as it does in the millikenvin range. It would have been observed as the failure of particles to block, down to the lowest temperatures, owing to quantum transitions. Then one would expect the Curie law, $M \propto 1/T$, to apply for the ZFC curve at all temperatures, rather than the typical blocking curve observed. There may be some non-trivial possibilities to reconcile blocking at high temperatures with tunneling at low temperatures. The ZFC curve shown in Fig. 5.18, and similar curves obtained by Awschalom and co-workers, were taken from undiluted samples. The authors of [79] then suggested that these curves manifest a spin-glass-like transition due to the interaction between particles, rather than one-particle blocking due to anisotropy barriers. In our opinion, this is unlikely. Spin-glass ordering due to the dipole interaction between particles is difficult to observe even in a system of ferromagnetic particles [118]. In antiferromagnetic particles, such as ferritin, the dipole interaction is too weak to provide ordering in the kelvin range. The value of the blocking temperature is consistent with the magnetic anisotropy expected in ferritin (see Chapter 5). The ZFC curves observed for ferritin, therefore, must be due to the ordinary blocking that has been observed in all particulate media [118].

Another possibility to resolve the contradiction between blocking at high temperature and tunneling at low temperature is to assume that the quantum transition rate is a non-monotonic function of temperature. This could be the case if the dissipation were to increase rapidly when the system is warmed up from millikelvin to kelvin temperatures, freezing the tunneling variable in one of the wells. One may expect this to happen owing to, e.g., the unfreezing of magnons and other modes with temperature. The third possibility to explain the disagreement between the high- and low-temperature behaviors arises from the difference between the magnitudes of the DC fields used in the ZFC and resonance experiments. The ZFC curve was obtained in a field of order 50 G, whereas the MQC experiment was performed in a record low field of 10^{-5} G. Coherent tunneling between the ground state levels takes place when the Zeeman term does not exceed the tunneling splitting. A higher field tunes the system off resonance, as is the case for $Mn_{12}Ac$ (see Chapter 7). Experimental evidence of this scenario was obtained in measurements of the field dependences both of the blocking temperature, T_B, and of the magnetic viscosity, S, [119]. Normally one should expect T_B to decrease and S to increase with increasing field because the field lowers the energy barriers. For ferritin, however, the two dependences $T_B(H)$ and $S(H)$ are different. In the case of $T_B(H)$, see Fig. 6.9, it first increases until about 2.5 kOe and then decreases at higher fields. This behavior of T_B with H has also been observed by Gider et al. [120], by Friedman et al. [121], and by Sappey et al. [122]. The fact that the maximum in the ZFC magnetization curve for ferritin is determined by the blocking of individual particles has been confirmed by AC measurements. It has been demonstrated that T_B depends logarithmically on the frequency of the AC field, the dependence expected from the relation between T_B and the timescale of the relaxation measurement. The $S(H)$ dependence at different temperatures is shown in Fig. 6.10. At 2.4 and 3 K it increases monotonically with the field. Between 3 and 8 K the viscosity first drops as the field increases from zero, then reaches a minimum at a cer-

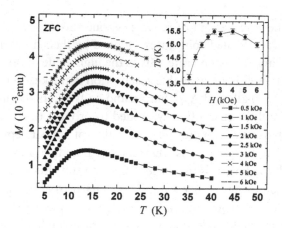

Figure 6.9 Zero-field-cooled magnetization curves obtained by cooling the ferritin sample in zero field and then heating it under various applied fields. The inset shows the field dependence of the blocking temperature. (Data from [119].)

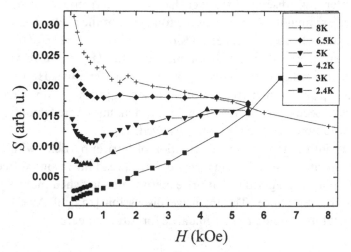

Figure 6.10 The magnetic viscosity as a function of the magnetic field at various temperatures. (Data from [119].)

tain field that depends on the temperature, and then increases at higher fields. Thus, in a certain temperature range, the system relaxes faster in a zero field than it does in a non-zero field.

The explanation suggested in [119] for the non-monotonic behavior of $T_B(H)$ and $S(H)$ in ferritin is that, at zero field, spin levels of each ferritin particle are doubly degenerate with respect to the spin reversal. This results in the resonant spin tunneling between degenerate pairs of levels. The field drives the particles away from the resonance, so that the tunneling rate decreases and T_B increases. At high field this effect is outweighed by the fact that the field also lowers the barrier, which leads to the decrease of T_B with the field.

The characteristic distance between the spin levels in antiferromagnetic ferritin must be of order $(\varepsilon_{an}\varepsilon_{ex})^{1/2}$, where ε_{an} and ε_{ex} are the anisotropy and exchange energies per atom. This gives about 10 K. To explain the high value of the field needed to tune the particles out of resonance, one should assume that tunneling takes place between excited spin states widened by their decay to the ground state and by their large tunneling rate. Given the high value of the crossover temperature in ferritin, this is a reasonable assumption in the kelvin range.

Equations (3.66) and (3.67) allowed Awschalom and co-workers to estimate the magnetic anisotropy and the susceptibility needed to provide the observed tunneling rate and the crossover temperature. The value of the susceptibility extracted in that way, $\chi_\perp \simeq 5 \times 10^{-5}$, is reasonable, but the anisotropy, $K \simeq 10^3$ erg cm^{-3}, is much too low. It is two orders of magnitude lower than the value obtained from the blocking temperature. This is hard to explain, given that the anisotropy of small particles is usually higher, not lower, than that of bulk materials, and that it is usually independent of temperature at low temperatures. In Chapter 3 we discussed the fact that the antiferromagnetic dynamics switches to the ferromagnetic one when the total spin of the particle, due to the non-compensation of sublattices, becomes of order $|S_1 - S_2| \simeq (H_a/H_{ex})^{1/2} S_1$. For the anisotropy, $K \simeq 10^5$ erg cm^{-3}, obtained from the ZFC curve, this limiting value is greater than the non-compensation in ferritin. However, at $K \simeq 10^3$ erg cm^{-3}, the 1% non-compensation is too high for the antiferromagnetic dynamics to occur. On the other hand, if the dynamics were ferromagnetic, the tunneling rate would be too low to be observed.

Finally, the width of the volume distribution of ferritin particles within each size group was greater than 10%. This is enough to spread the resonant frequencies over one order of magnitude. Future experiments may shed more light on the nature of the resonance. The intensive discussion [123] of Awschalom's work in the literature should provide guidance for future researchers.

6.3   The motion of domain walls in mesoscopic wires

An attempt to observe tunneling of domain walls was made by Hong and Giordano [15]. They chose for their experiment a mesoscopic nickel wire of length 10–30 μm and diameter less than 35 nm. The direct measurement of the magnetization of such a wire is difficult. Instead, they measured the magnetoresistance of the wire, Fig. 6.11.

Curves 1 and 2 in Fig. 6.11 show the magnetoresistance when the field was swept up and down, respectively. Without saying much about the mechanism of the magnetoresistance, the authors attributed the hysteresis to the metastable magnetic states arising from the pinning of domain walls inside the wire. This suggestion is supported by the fact that $R(H)$ is completely reversible if, moving up along curve 1, one stops at 100 Oe and sweeps the magnetic field back down. Likewise, $R(H)$ on the way down is completely reversible if one does not go below -100 Oe along curve 2. It is then reasonable to assume that the defects of

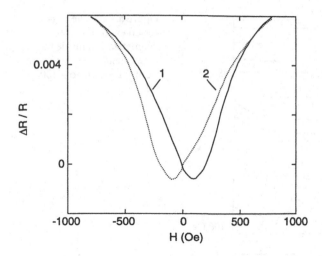

the magnetic structure (domain walls), which can be pinned by the imperfections in the wire, exist inside the wire for -100 Oe $< H < 100$ Oe. Since electrons are scattered by domain walls, it is then clear that the change in the magnetoresistance should have its maximum inside this field range. The question of what causes the rapid change in the magnetoresistance outside this range, however, remains unanswered. In our opinion, this change occurs owing to the progressive inhomogeneity of the magnetization inside the wire when the field is reduced, Fig. 6.12.

The more inhomogeneous the magnetization inside the wire the more spin scattering of the electrons occurs. Clearly, in the absence of pinning centers, the most inhomogeneous distribution of the magnetization, corresponding to zero total magnetic moment of the wire, would occur at $H = 0$. In this case the extremum of $R(H)$ would also occur at $H = 0$. However, in the presence of pinning centers, zero total moment occurs at the depinning field, which for the nickel wire in question apparently was about 100 Oe at 4.2 K. Figure 6.11, therefore, must resemble the hysteresis curve of the wire, which is difficult to obtain directly. This part of the conclusions of Hong and Giordano leaves little doubt.

Let us now turn to their conclusions regarding the evidence for domain wall tunneling. It is known that the area of the hysteresis loop obtained at a fixed sweeping rate depends on the temperature. In a simple picture of the depinning of a domain wall this can be explained as follows. Let the sweeping rate be γ s^{-1}, the energy barrier associated with the depinning be $U(H)$, and the thermal depinning rate be $\Gamma = \omega_0 \exp\left[-U(H)/T\right]$. Then the effective depinning field, $H^*(T)$, can be obtained from the condition $\gamma = \Gamma$. The depinning of a domain wall of small area from a planar defect has been studied in [42]. The barrier disappears at some field H_0. Close to that field it behaves as $(H_0 - H)^{3/2}$. Consequently, the effective depinning field must behave as $H^* = H_0 - CT^{3/2}$, where C is a constant that depends on the sweeping rate. The field H_0 then gives a physical depin-

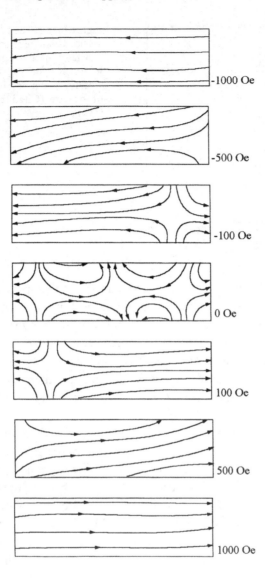

Figure 6.12 The suggested evolution of the magnetization inside the wire when the field is swept from −1000 Oe to +1000 Oe.

-1000 Oe

-500 Oe

-100 Oe

0 Oe

100 Oe

500 Oe

1000 Oe

ning field at $T = 0$ that does not depend on the sweeping rate. The actual $H^*(T)$ dependence obtained in experiment is shown in Fig. 6.13. The derivative of the $H^*(T)$ curve seems to go to zero as the temperature is lowered, whereas the theory predicts a totally different behavior if the depinning is due to thermal over-barrier transitions. If, however, quantum depinning is involved, one should replace T in the formula for $H^*(T)$ by $T_{\text{esc}}(T)$ discussed in Chapter 2. This would give a zero derivative in the quantum regime. Hong and Giordano considered this fact to be evidence of quantum tunneling. We should notice, however, that it relies heavily upon the theoretical dependence of the pinning barrier on

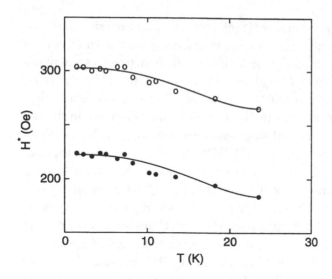

Figure 6.13 The temperature dependence of the effective depinning field for the 23 nm Ni wire. (Cited from [15].)

the magnetic field. That dependence was obtained for a very simplified special case of depinning from a planar defect, which is unlikely to be the case if the defect is not manufactured artificially. If the dependence of the barrier on the field is $(H_0 - H)^\beta$, with $\beta < 1$, the low-temperature behavior of $H^*(T)$ owing to thermal transitions alone would be qualitatively of the form observed in the experiment. Despite that current uncertainty, we should say that the method suggested by Giordano and collaborators is very promising for observing tunneling of the magnetization.

6.4   Experiments on single particles

Most of the classical theory of monodomain magnetic particles was developed during the 1950s and 1960s [26,124]. Until very recently, however, like the theory of spin tunneling, it had never been tested directly in experiments. The reason it took so long is that, to test the predictions of the micromagnetic theory quantitatively, one has to perform measurements on a single particle, which requires an extremely high sensitivity. To avoid this problem one could perform measurements on an ensemble of identical particles, but the task of manufacturing atomically and magnetically identical monodomain particles is even more difficult. Measurements of single monodomain particles, as small as 15 nm in diameter, have recently become possible [125] owing to progress in the lithographic fabrication of nanoscale-sized particles and in manufacturing extremely sensitive micro-SQUIDs.

Wernsdorfer *et al.* [14,126] studied magnetization reversal in single nanoparticles and nanowires of Co and Ni. Nearly spherical Co particles of diameter 15–30 nm were obtained by the method of discharging an arc into carbon nanotubes [126]. According to the authors of [126], these particles were single crystals with a surface roughness of about two atomic layers, encapsulated within carbon which protected them from oxidation. In a different method, particles and nanowires of diameter 30–100 nm and length up to 5 μm were obtained by electrodeposition in nanoporous polycarbonate membranes. The membrane was then dissolved in chloroform. In both methods the particles were dispersed in ethanol by ultrasonification. A drop of that suspension was placed on a chip of about 100 SQUIDs. Planar micro-bridge DC SQUIDs were used. They were made of Nb, which insured their operation at temperatures up to 7 K. To prevent the magnetic flux from being trapped in the SQUIDs, the SQUIDs were made of a single Nb layer of thickness 20 nm. After the ethanol containing the magnetic particles had evaporated, the particles stuck to the chip owing to molecular forces. The magnetic signal was sufficient for taking measurements for nanoparticles that had been left on the micro-bridge of a SQUID. Because of the close proximity of the particle and the SQUID, it was possible to detect changes in magnetization by as little as $10^4 \mu_B$. The actual position and shape of the particle were determined by scanning electron microscopy after the magnetic measurements had been taken.

A magnetic field up to 1 T was applied in the plane of the SQUID and the angular dependence of the switching field H_{sw} (that is, the field at which magnetization reversal occurs) was measured as a function of the direction of the field, θ. For Ni wires, the dependence $H_{sw}(\theta)$ exhibits a crossover between two different modes of magnetization reversal for wires of diameter about 80 nm, Fig. 6.14. The dotted line in Fig. 6.14(a) is the prediction of the curling model for an infinite cylinder [14]. The transition from the curling mode to the uniform rotation, when the diameter of the wire decreases below 80 nm, is in accordance with our expectation from the micromagnetic theory. Indeed, owing to the large exchange energy, the curling can develop on a spatial scale comparable to the width of a domain wall. As soon as one goes below that scale, uniform rotation of the magnetization should be expected.

The smallest particles studied to date were nearly spherical monocrystalline Co particles of diameter 15–30 nm and surface roughness about two atomic layers. Figure 6.15 shows the typical angular dependence of the switching field in these particles. The maximum at $\theta = 0$ clearly indicates the tendency toward uniform rotation. One should notice, however, that, in theory, if uniform rotation were to take place, the height of that maximum should be the same as that at $\pm 90°$, which is also true for the Ni wires discussed above. The origin of the discrepancy could be twofold. First, the size of the Ni and Co particles studied in these experiments appears to be close to the size for crossover from curling to uniform rotation. It is possible, therefore, that the height of the $\theta = 0$ maximum will increase when smaller particles are measured. Secondly, observation

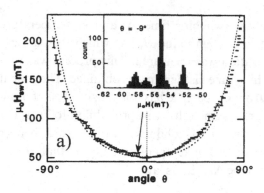

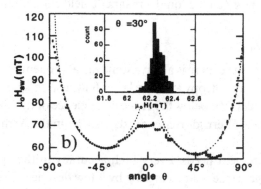

Figure 6.14 The angular variation of the switching field in Ni nanowires: (a) for wires of diameter 92 nm and length 5 μm, and, (b) for wires of diameter 50 nm and length 3.5 μm. (Cited from [14].)

of purely uniform rotation in particles of that size can be hampered by the presence of extended defects.

Let us now turn to the dynamics of the magnetization reversal observed in these smallest Co particles. Two techniques were used by Wernsdorfer *et al.* to study the probability of the switching in the presence of a finite energy barrier.

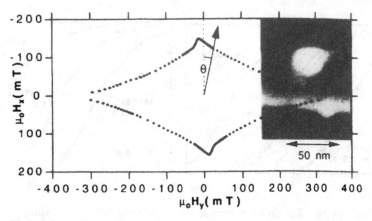

Figure 6.15 The angular dependence of the switching field in Co nanoparticles. (Cited from [126].)

The first was measurement of the switching time. At a given temperature, the magnetic field was increased to a given value H_w, at which the elapsed time needed to switch the magnetization was measured. This process was repeated about 100 times and the switching time histogram was obtained. The probability that the magnetization of a 20 nm Co particle has not yet switched at a time t is shown in Fig. 6.16. The authors assumed that this probability follows a simple statistical distribution: $P(t) = \exp(-t/\tau)$ (solid lines in Fig. 6.16), τ being a single lifetime of the metastable state. For temperatures higher than 0.3 K they found that the lifetime followed the Arrhénius law

$$\tau(T, H) = \tau_0 \exp[U(H_w)/T], \qquad (6.10)$$

where the dependence of the energy barrier on the magnetic field can be approximated by

$$U(H) = U_0(1 - H/H_0)^{\alpha} \qquad (6.11)$$

with the coefficient α close to $\frac{3}{2}$. Since there is always some misalignment between the direction of the field and the anisotropy axis, the latter finding is in agreement with theory (see Chapter 3, model IV). Below 0.3 K the dependence of the lifetime on the temperature, according to Wernsdorfer et al., departs from the Arrhénius law behavior.

The switching time measurement needs a very high accuracy in the applied field, since even a small change in the field changes τ by a few decades. For that reason, another method, namely switching field measurement, has also been used to study the field dependence of the lifetime. In that method, at a fixed temperature, the applied field was changed at a given rate. The value of the field that switched the magnetic moment of the particle was recorded. The sweeping rates used were in the range 0.01–100 mT s^{-1} for temperatures in the range 0.1–6 K. The switching field histograms were obtained from about 100 cycles.

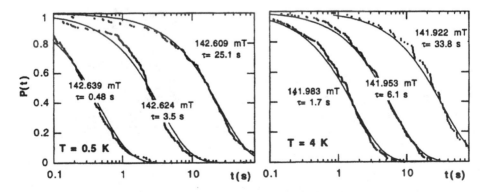

Figure 6.16 The time dependence of the probability of 'non-switching' of a 20 nm Co particle for various applied fields and temperatures. Solid lines represent exponential fits to the experimental curves. (Cited from [126].)

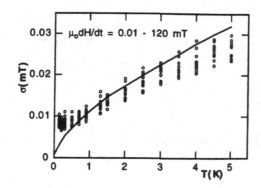

Figure 6.17 The width of the switching field distribution versus temperature. (Cited from [126].)

From these histograms the mean switching field, H_{sw}, and the width of the distribution, σ, were determined. If the switching were purely thermal, the width of the switching field distribution would go to zero in the limit of zero temperature. Indeed, in that case, the switching at $T = 0$ would occur at a field determined by $U(H) = 0$. The dependence of σ on temperature is shown in Fig. 6.17. It indicates a possible non-zero probability of switching in the limit of zero temperature and a finite barrier. The authors of [14] and [126] suggested quantum tunneling of the magnetization as a possible explanation for their data.

Chapter 7

Tunneling in magnetic molecules

Magnetic molecules represent another interesting area for MQT studies. In this chapter we will describe experiments on spin tunneling in macroscopic crystals of $Mn_{12}Ac$ molecules. A $Mn_{12}Ac$ molecule has spin 10 and is equivalent to a very small single-domain magnetic particle. A crystal consists of a macroscopic number of molecules arranged in a tetragonal lattice. Spins of the molecules interact weakly with each other and are subject to a very strong uniaxial anisotropy. A $Mn_{12}Ac$ crystal is, therefore, equivalent to a large system of identical uniformly oriented magnetic particles, which has been wanted so much by experimentalists. Experiments on $Mn_{12}Ac$ [16] are the first experiments on spin tunneling which have been understood quantitatively without any free parameters. Whether the moment of the molecule, $20\mu_B$, is macroscopic enough to allow one to talk about MQT is left to the judgement of the reader.

7.1   The Mn-12 acetate complex

The chemical formula of the manganese acetate complex discussed below is $[Mn_{12}O_{12}(CH_3COO)_{16}(H_2O)_4]\cdot 2CH_3COOH\cdot 4H_2O$. We will denote it by $Mn_{12}Ac$. This compound was first synthesized by Liz and characterized by X-ray diffraction methods [128]. It forms a molecular crystal of tetragonal symmetry with the lattice parameters $a = 1.732$ nm and $b = 1.239$ nm. The unit cell contains two $Mn_{12}O_{12}$ molecules surrounded by four water molecules and two acetic acid molecules. The structure of the $Mn_{12}O_{12}$ molecule is shown in Fig. 7.1. The molecule possesses S_4 symmetry. It has a tetrahedral core of four Mn^{4+} ions at its center, surrounded by a crown of eight Mn^{3+} ions. Each of the Mn^{3+}

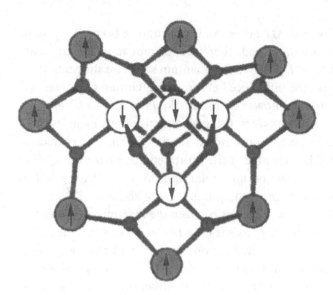

Figure 7.1 The structure of the $Mn_{12}O_{12}$ molecule. Big dark circles are Mn^{3+} ions. Big white circles are Mn^{4+} ions. Small circles are O^- ions.

ions has spin $\frac{3}{2}$, whereas each of the Mn^{4+} ions has spin 2. The spins of the Mn ions are coupled through oxygen ligands which allow an indirect exchange interaction between them. Owing to the acetate shell, each molecule is well isolated from the others. The magnetism of the molecule is of purely spin origin; the orbital moment is quenched by the crystal field. This is confirmed by the value of the g-factor, $g = 1.9$, obtained from ESR studies [129]. Low-field magnetic measurements [130] (see below) show that the total spin of the molecule is 10. This can be understood as resulting from the ground state with eight ions of the crown forming a ferromagnetic sublattice of spin $8 \times 2 = 16$, four ions of the core forming a ferromagnetic sublattice of spin $4 \times \frac{3}{2} = 6$, and the two sublattices being antiparallel, $S = 16 - 6 = 10$. This magnetic structure of the molecule apparently persists well above 50 K, indicating that the exchange interaction is strong enough for the molecule to be treated in the kelvin range as a nanomagnet of a rigid spin 10.

The symmetry of the molecule and of the lattice results in strong uniaxial anisotropy along the c-axis. Thus, to a first approximation, the Hamiltonian of the $Mn_{12}O_2$ nanomagnet must be

$$\mathcal{H} = -DS_z^2 - g\mu_B \mathbf{S} \cdot \mathbf{H}, \tag{7.1}$$

where the z-direction is chosen to be along the c-axis. This can be tested by applying a transverse field, $\mathbf{H} \perp c$, and measuring the magnetization curve. It should exhibit no hysteresis. As $H_\perp$ increases, spins must gradually rotate toward the direction of the field. For the classical magnetic moment, one can easily work out from Eq. (7.1) that the magnetization must attain saturation, $M_s = g\mu_B S$, at the anisotropy field, $H_a = 2DS/(g\mu_B)$, and must be linear with respect to $H_\perp$

below H_a. For the quantum spin, $M_\perp$ never attains saturation because $S_\perp$ does not commute with the Hamiltonian and, therefore, cannot have a particular value $S_\perp = S$. Theoretical $M_\perp(H_\perp)$ curves for quantum spins are shown in Fig. 7.2. For a spin as large as 10, the difference between the quantum and classical cases is already negligible, which allows one to test Eq. (7.1) and to obtain the value of the anisotropy field. The transversal magnetization curve for single-crystalline $Mn_{12}Ac$ is shown in Fig. 7.3. It saturates at about 9 T and, as expected, is linear with respect to the field. The value of the saturated magnetization agrees well with the value expected from the molecular spin, $S = 10$. With $g = 1.9$ obtained from the ESR measurements [129], this gives D of about 0.6 K.

In the absence of a field, the energy barrier between the two opposite orientations of S with respect to the anisotropy axis is $U = DS^2 = 60$ K. Consequently, on the timescale of a typical ZFC magnetization measurement the transitions between the two orientations should be blocked below 3 K. This is in accordance with experiment, as illustrated by Fig. 7.4. Note that earlier reports on the magnetism of $Mn_{12}Ac$ interpreted the peak in the ZFC magnetization curve as a phase transition owing to a change in spin order inside the molecule. This is apparently not true, as one can see from the specific-heat measurements, Fig. 7.5. The $c(T)$ curve [131] shows no anomaly at 3 K that would indicate a phase transition. Given that the value of the blocking temperature agrees with the

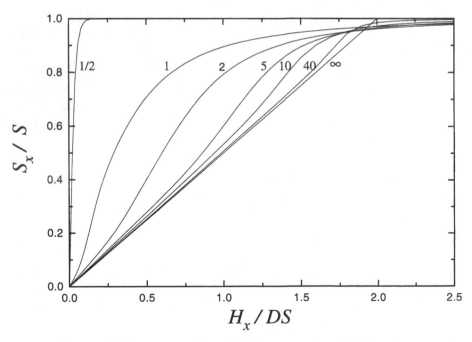

Figure 7.2 The transversal magnetization curve of a uniaxial quantum magnet with Hamiltonian $\mathcal{H} = -DS_z^2 - \mu_B S \cdot H$ for various values of S.

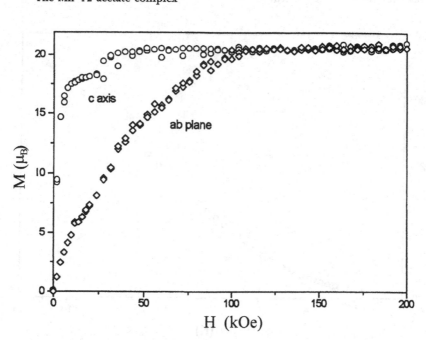

Figure 7.3 The magnetization curve in the c-direction and in the a–b-plane for the Mn$_{12}$Ac crystal. (Cited from [131].)

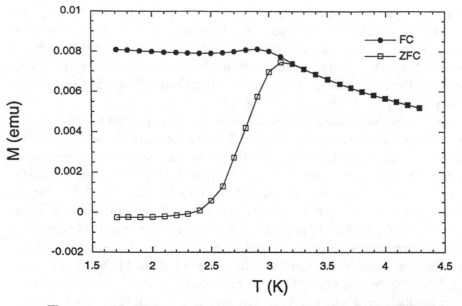

Figure 7.4 The ZFC magnetization curve of the Mn$_{12}$O$_{12}$ crystal. (Cited from [16].)

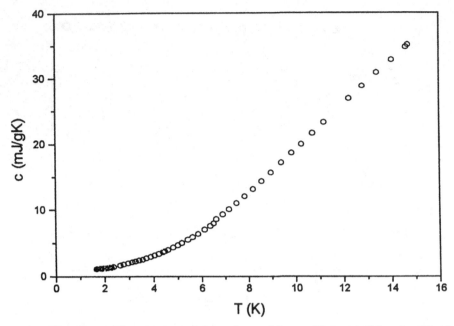

Figure 7.5   The temperature dependence of the specific heat of $Mn_{12}Ac$. (Cited from [131].)

measured value of the anisotropy barrier, the peak at 3 K in the ZFC magnetization curve should be attributed to the transition from the blocked state at $T < 3$ K to superparamagnetic behavior at $T > 3$ K. This, together with the value of the spin, is further confirmed by the fact that, above the blocking temperature, the susceptibility of the crystal is described well by the Curie–Weiss law, $\chi = (g\mu_B S)^2/T$, with $S = 10$, Fig. 7.6 [132].

Let us now discuss other interactions that may be important for understanding the magnetic behavior of $Mn_{12}Ac$. The magnetic dipole field exerted by other molecules on any given molecule depends on the local arrangement of spins in the crystal. A typical value of the dipole field is of the order of 100 G. There is also a comparable magnetic field exerted on the molecular spin by nuclear moments owing to the hyperfine interaction. Each Mn nucleus has spin $\frac{5}{2}$, which exerts a field of about 250 G on the ion [133]. The resulting field exerted on the molecule depends on the mutual orientation of the nuclear spins, which is random. The maximal value of that field has been estimated to be around 500 G [133], but the average should be in the range 100–200 G. We should remember these numbers since both the dipole field and the hyperfine field have transverse components. In the absence of an external $H_\perp$, these are the terms in the Hamiltonian which violate its commutation with S_z, leading to the relaxation of M_z toward thermodynamic equilibrium. The tetragonal symmetry of the molecule must result in the anisotropy term, $C(S_+^4 + S_-^4)$, in the Hamiltonian. The

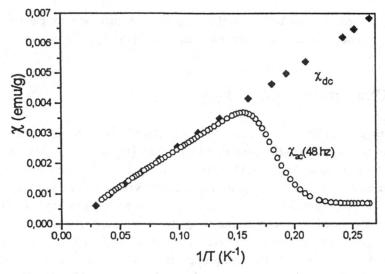

Figure 7.6   The magnetic susceptibility of $Mn_{12}Ac$ versus $1/T$. (Cited from [131].)

constant C is difficult to compute from first principles. It is clear, however, that the tetragonal anisotropy should be very weak compared with the uniaxial anisotropy. Finally, there should be terms in the Hamiltonian responsible for spin–phonon interaction. At zero temperature their contribution can be absorbed into the anisotropy terms. At finite temperature these terms must be responsible for activating some molecules to excited spin states. These states are shown in Fig. 7.7. They are classified in terms of the $m = S_z$ quantum number. According to Eq. (7.1), there are $2S + 1 = 21$ energy levels $E_m = -Dm^2 - g\mu_B H_\parallel m$, with $m = -10, -9, ..., 0, ..., 9, 10$. The hyperfine interactions split each m-level into 861 sublevels [133]. The distance between the sublevels is so small that it is reasonable to talk about continuous energy bands of width $\Delta \approx 10^{-2}$ K originating from each m-level. Interactions that do not commute with S_z cause transitions between different bands inside the well and between the two wells shown in Fig. 7.7. One

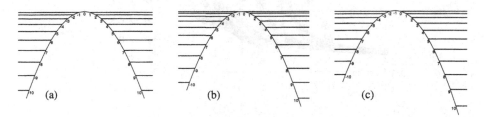

Figure 7.7   Energy levels of $Mn_{12}O_{12}$ molecules in a field applied along the anisotropy axis: (a) $H_\parallel = 0$, (b) $H_\parallel = H_a/(2S)$, and (c) $H_\parallel = H_a/S$.

amazing feature of $Mn_{12}Ac$ is that these quantum transitions are clearly evident in macroscopic magnetization measurements [16, 134–136].

7.2   Quantum magnetic hysteresis

Unusual magnetic hysteresis curves for $Mn_{12}Ac$ samples had been published by a number of researchers [132] before Friedman *et al.* [16] observed the regularly spaced *jumps* in the magnetization curve and recognized that they carried the key to the understanding of the quantum mechanics of this compound. Typical hysteresis loops obtained at various temperatures are shown in Fig. 7.8. These loops were obtained by sweeping the magnetic field at a constant rate $r = 60$ mT min^{-1}. The steps in the magnetization are evident when the field is in the direction of M but not when the field is opposite to M. Let us now try to understand these findings by applying what we already know about $Mn_{12}Ac$. The magnetic field along the anisotropy axis could not change the magnetization of the system if spins of individual $Mn_{12}O_{12}$ molecules were frozen in their initial directions. Owing to the finite-anisotropy barrier, however, there are transitions between two opposite orientations of spins with respect to the anisotropy axis.

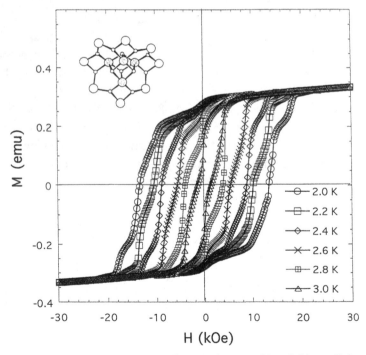

Figure 7.8   Hysteresis loops of a $Mn_{12}Ac$ crystal in a field parallel to the anisotropy axis. (Cited from [16].)

If the sweeping rate of the field were infinitely slow, one would obtain a reversible magnetization curve, corresponding to the thermal equilibrium at any given value of the field, Fig. 7.9(a). In contrast, if the sweeping rate were infinitely fast, the spin of the individual molecule would switch to the opposite orientation only at $H = H_a$, at which the barrier disappears. This would lead to the square hysteresis loop shown in Fig. 7.9(b).

The real magnetization curves (Fig. 7.8) are intermediate between the two extremal curves shown in Fig 7.9. They are governed by the detailed balance between the 21 m-levels of spin 10. As a very crude approximation one can write for the total moment

$$\frac{\mathrm{d}M}{\mathrm{d}t} = -\Gamma(H)[M(t) - M_{\mathrm{eq}}(H)], \tag{7.2}$$

where Γ is the rate of relaxation of M toward the equilibrium value M_{eq}. The dependence of Γ on H can be extracted from the magnetization curve if one notices that, at a given sweeping rate, $r = \mathrm{d}H/\mathrm{d}t$, the time derivative of M equals $(\mathrm{d}M/\mathrm{d}H)/r$ [137]. This gives

$$\Gamma(H) = \frac{r(\mathrm{d}M/\mathrm{d}H)}{M_{\mathrm{eq}} - M(H)}. \tag{7.3}$$

The field dependence of the rate calculated from the magnetization data is shown in Fig. 7.10. It reveals striking maxima at temperature-independent values of the longitudinal field, satisfying

$$H_{\parallel} = H_n = nH_0, \tag{7.4}$$

with $n = 0, 1, 2, \ldots$ and $H_0 \approx 0.46$ T. The maxima exhibit a Lorentzian shape, as is illustrated by Fig. 7.11.

Let us now try to explain these maxima on the basis of our knowledge of the magnetic structure of $Mn_{12}Ac$. Imagine that the crystal had been magnetized in

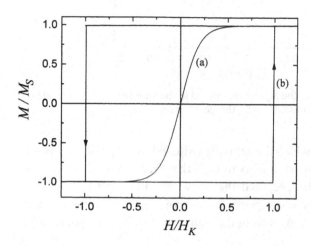

Figure 7.9 Hypothetical hysteresis loops: (a) at an infinitely slow sweeping rate and (b) at an infinitely fast sweeping rate.

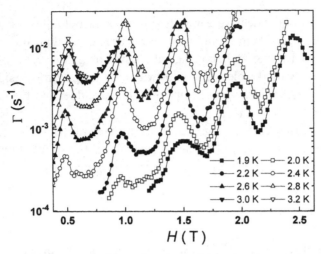

Figure 7.10   The field dependence of the relaxation rate in a $Mn_{12}Ac$ crystal, calculated from the initial magnetization curves, using Eq. (7.3). (Data from [135].)

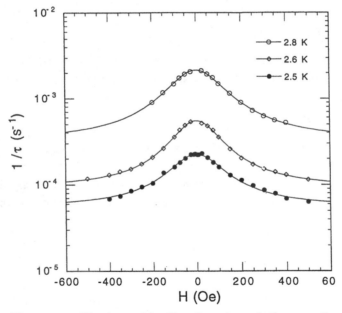

Figure 7.11   The shape of the $H_\parallel = 0$ maximum in the magnetic relaxation rate of $Mn_{12}Ac$. The solid line shows the fit by the Lorentzian.

the negative z-direction, after which the field was switched to the positive z-direction and the moment was allowed to relax toward the direction of the field. The m-levels of $Mn_{12}O_{12}$ molecules prepared that way are illustrated by Fig. 7.7. The levels in the left-hand well correspond to the projections of S_z against the direction of the field, whereas the levels in the right-hand well correspond to S_z

in the direction of the field. At the beginning of the relaxation process the molecules are thermally distributed over the levels in the left-hand well. At certain values of the field, the m-levels in the left-hand well match the m-levels in the right-hand well. At these field values resonant tunneling between the two wells takes place, leading to the maxima in the relaxation rate. Let us show that this picture is in quantitative agreement with experiment [16,134]. According to Eq. (7.1), the condition that two levels, m and m', match, is

$$-Dm^2 - g\mu_B H_\parallel m = -Dm'^2 - g\mu_B H_\parallel m'. \tag{7.5}$$

This gives $m + m' = -g\mu_B H_\parallel / D$, which can be satisfied at the values of the field, $H_\parallel = nH_0$, with $n = 0, 1, ..., 20$ and

$$H_0 = \frac{D}{g\mu_B} = \frac{H_a}{2S}. \tag{7.6}$$

Given the values of $H_a = 9.2$ T and $S = 10$, measured independently, this is in excellent agreement with the resonant field values observed in experiments, Fig. 7.12. The first three resonances, at $n = 0$, $n = 1$, and $n = 2$, are shown in Fig. 7.7. An additional important observation is that, at $H_\parallel = nH_0$, each m-level with $m \geq 10 - n$ in the right-hand well matches the $m' = n - m$ level in the left-hand well, leading to the multiple level crossing in the two wells. At $n = 0$ this is the consequence of the time-reversal symmetry. At $n > 0$ it has to do with the fact that the Hamiltonian (7.1) contains only first- and second-order terms in S_z.

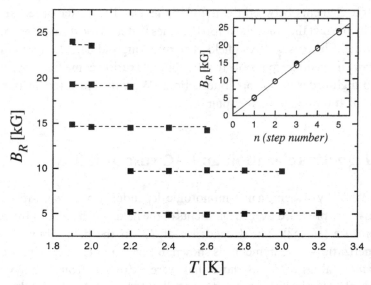

Figure 7.12 Values of the first five resonance fields obtained at various temperatures. The insert shows that they are equally spaced.

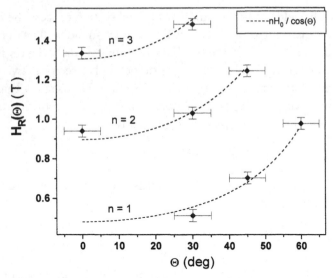

Figure 7.13 The dependence of the resonant fields $H_{1,2,3}$ on the orientation of the field. Dashed lines show the fit by $H_n = nH_0/\cos\theta$. (Cited from [135].)

Further support for the above picture comes from the magnetization measurements when the field is applied at an angle θ to the anisotropy axis. The positions of the maxima in the relaxation rate exhibit a $1/\cos\theta$ dependence on the angle, shown in Fig. 7.13. This suggests, in accordance with Eq. (7.5), that the resonance condition is determined by the longitudinal component of the field, whereas the transversal component provides the term in the Hamiltonian that has non-vanishing matrix elements between the matching levels. This latter observation is supported by the fact that the relaxation becomes faster as the transverse component of the field increases. If the field is purely longitudinal, the terms in the Hamiltonian which are responsible for the spin relaxation come from hyperfine, dipole, and high-order anisotropy interactions. We will return to this question in more detail at the end of this chapter.

7.3 Magnetic relaxation and AC susceptibility

Another possibility of seeing a non-monotonic dependence of the spin-relaxation rate in $Mn_{12}Ac$ is provided by measurements of the field dependence of the blocking temperature [16,134]. The dependence of T_B on the field, obtained from ZFC magnetization measurements, is shown in Fig. 7.14. The maxima occur at the very same values of $H_\parallel$ as those that were extracted from the hysteresis loops. From the classical perspective this oscillating behavior of T_B with respect to H would be difficult to understand. Indeed, if T_B were given by Eq. (5.13), it

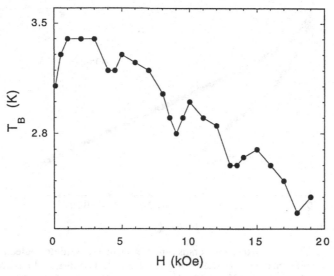

Figure 7.14 The field dependence of the blocking temperature in $Mn_{12}Ac$. (Cited from [16].)

would decrease monotonically with increasing field since the field reduces the energy barrier between the two orientations of S along the anisotropy axis. The non-monotonic dependence of T_B on H indicates that the field exerts a more subtle effect, which should be understood along the lines discussed in the previous section.

Up to now we have described experiments in which the relaxation rate in $Mn_{12}Ac$ has been measured indirectly, on the basis of theoretical assumptions about the relation between the relaxation and the magnetization curve. The direct measurement of the temperature and field dependences of the relaxation is more demanding insofar as it requires hours of experimental time to obtain each experimental data point. For $Mn_{12}Ac$ such measurements have been performed in great detail. A typical time dependence of the magnetic moment is shown in Fig. 7.15. A few observations are in order. First, the relaxation sometimes appears to be slower in a stronger field, as is illustrated by Fig. 7.15. Secondly, after some initial waiting period, it becomes exponential in time [16],

$$M(t) = M_{eq}(H)\{1 - \exp[-\Gamma(H)t]\}, \qquad (7.7)$$

as expected for a set of identical $Mn_{12}O_{12}$ particles characterized by a single energy barrier. In fact, the entire time dependence of the magnetization that includes short times alongside long times can be fitted by two exponents, having different relaxation times. This has been observed by all groups working on different samples [16, 134–136] and, apparently, is an intrinsic property of

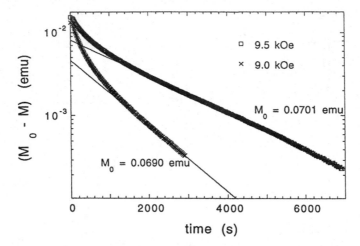

Figure 7.15   The magnetization versus time for a $Mn_{12}Ac$ sample cooled to 2.4 K in zero field and then exposed to fields of 0.9 and 0.95 T. The straight lines are fits to an exponential function after roughly the first 10^3 s. (Cited from [16].)

the system. We shall assume that there are two magnetic 'species' responsible for these different relaxation constants and see what can be extracted from experimental data about these two 'species'. The dependence of the long-time relaxation rate, Γ, on H obtained directly from the relaxation measurements is shown in Fig. 7.16. The maxima of the rate are again temperature-independent and

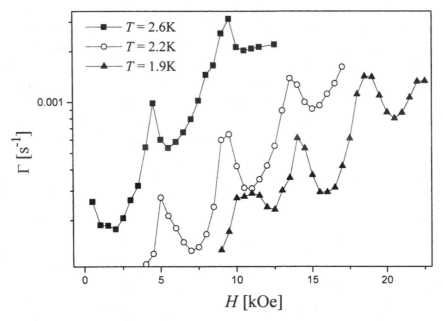

Figure 7.16   The relaxation rate in $Mn_{12}Ac$ obtained from relaxation experiments at three different temperatures. (Cited from [135].)

occur at exactly the same values of the field as those obtained from the $M(H)$ measurements. This is a remarkable observation given that Γ obtained from the relaxation measurement and Γ extracted from $M(H)$ are different. They correspond to the two different 'species' discussed above. The latter is the consequence of the fact that the sweeping rate used to obtain the hysteresis loops is faster than the relaxation rate. Consequently, measurement of the hysteresis loop probes the short-time regime shown in Fig. 7.15. The fact that the relaxation peaks at the same values of the magnetic field in both regimes suggests that the two 'species' are $Mn_{12}O_{12}$ molecules subject to different transverse fields (dipole, hyperfine, etc.) that are causing the transitions.

Let us now turn to the AC susceptibility measurements. Although they do not bring qualitatively new information compared with the DC measurements, they can be more accurate and probe shorter time scales [135, 138]. For a single barrier, as is the case for $Mn_{12}Ac$, the in-phase and out-of-phase components of the AC susceptibility are given by formulas (6.3), χ_0 being the DC superparamagnetic susceptibility, $\chi_0 = (g\mu_B S)^2/T$. The temperature dependences of χ' and χ'' obtained at 5 Hz AC field are shown in Fig. 7.17. It exhibits the characteristic behavior of an ideal superparamagnetic system with a single energy barrier. At 5 Hz the blocking temperature is about 6 K, which is twice the T_B observed in the ZFC magnetization measurements, Fig. 7.4. This is what one should expect from the ratio $\ln(\nu t)/\ln(\nu/\omega)$ of the two blocking temperatures owing to the different timescales of the ZFC magnetization and AC susceptibility measurements. Above the blocking temperature $\omega\tau(T)$ rapidly becomes exponentially small, which gives $\chi'' \to 0$ and $\chi' \propto 1/T$, in accordance with the AC data shown in Fig. 7.17. According to Eq. (6.3), the maximum of χ'' with respect to the frequency should occur at $\omega = \tau^{-1} = \Gamma$. Thus, by sweeping the frequency at different temperatures and DC fields one can obtain the dependence of τ on T and H. The field dependence of the relaxation time obtained by this method is plotted

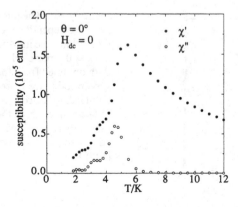

Figure 7.17 The temperature dependences of χ' and χ'' obtained at 5 Hz AC field for a $Mn_{12}Ac$ sample. (Cited from [135].)

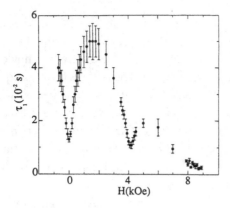

Figure 7.18 The field dependence of the relaxation time in $Mn_{12}Ac$ obtained from AC susceptibility measurements. (Cited from [138].)

in Fig. 7.18. It shows at least three minima (maxima of the relaxation rate) that occur at exactly the same values of the field as those obtained in the magnetization and relaxation measurements. At a fixed value of the field one can also study the temperature dependence of the relaxation time by this method. The corresponding data, collected both from AC susceptibility and from relaxation measurements at $H = 0$ are shown in Fig. 7.19. These data illustrate one point that is crucial for the understanding of the resonant magnetic phenomena in $Mn_{12}Ac$: The temperature dependence of the relaxation time follows the Arrhenius law, $\tau = \tau_0 \exp(U/T)$, the attempt time τ_0 being of the order of 10^{-7} s. The energy barrier, U, obtained from the slope in Fig. 7.19, is about 60 K at $H = 0$, in good agreement with the value of the barrier obtained from the measurement of the magnetic anisotropy. Both the relaxation and the AC susceptibility measurements also indicate that, as the field increases, the slope of the straight line in Fig. 7.19 decreases in accordance with the field dependence of the barrier, $U = U_0(1 - H_\parallel/H_a)^2$.

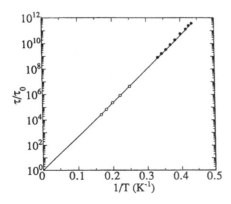

Figure 7.19 The temperature dependence of the relaxation time in $Mn_{12}Ac$ at $H = 0$: open circles are data obtained from the AC susceptibility measurements; closed circles are data obtained from the magnetic relaxation measurements. (Cited from [138].)

7.4   **The theory of thermally assisted spin tunneling in $Mn_{12}Ac$**

We shall try now to reconcile two experimental facts: (1) that the relaxation follows an Arrhénius law, which indicates its thermal nature; and (2) that the relaxation has maxima at fields corresponding to the level crossing, which must be of quantum origin. The natural explanation for these facts [16] comes if one assumes that the resonant tunneling responsible for the maxima is thermally assisted, as illustrated in Fig. 7.20. When a metastable magnetic state is prepared, spin levels in the left-hand well become thermally populated, with the population of each level inside the ensemble of $Mn_{12}O_{12}$ molecules being proportional to the Arrhénius factor $\exp(-E_m/T)$. At $H \neq H_n$, magnetic relaxation takes place via overbarrier thermal transitions into the right-hand well, presumably owing to the absorption and emission of phonons. Owing to the conservation of angular momentum, each phonon can change its m-number by one. Thermal relaxation then occurs via a staircase absorption of phonons on the way up in the left-hand well, followed by the emission of phonons on the way down in the right-hand well [131]. The thermal rate, of course, is given by the Arrhénius law and the actual mechanism only becomes important if one wants to compute the Arrhénius prefactor.

At $H_{\parallel} = H_n$ each m-level in the left-hand well matches the $(n - m)$-level in the right-hand well, and resonant tunneling takes place for each matching pair. The tunneling addition to the thermal activation apparently produces the maxima in the relaxation rate. Let us now discuss what interactions can be responsible for tunneling. First, we shall argue that the experimental data unambiguously impli-

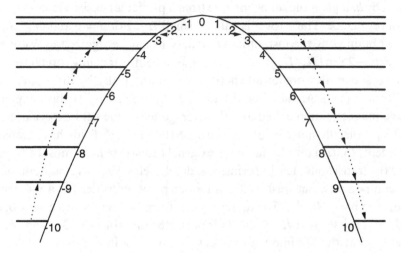

Figure 7.20   Thermally assisted resonant tunneling from excited levels in $Mn_{12}Ac$.

cate the non-diagonal interactions which are of first order in S_x and S_y, that is, field-induced ones. Indeed, let us assume that the corresponding interaction is of the form AS_x^p. At $p = 1$, it is just the Zeeman term, $-H_\perp \cdot S_\perp$. Interactions with $p = 2, 4, ...$ must be due to the particular structure of the magnetic anisotropy. Each order of the perturbation theory on AS_x^p changes the spin projection by p. Tunneling from m to $n - m$ will appear, therefore, in the $(2m - n)/p$th order of the perturbation theory. This can be satisfied by all $n = 0, 1, 2, ..., 20$ for $p = 1$ but not for $p > 1$. At, e.g., $p = 2$ tunneling is only possible at $n = 0, 2, 4, ..., 20$, that is, at values of $H_\parallel$ twice those given by Eq. (7.6). In other words, at $p = 2$ only every second resonance would result in a maximum in the relaxation rate, in apparent disagreement with the experimental data. Greater p will produce selection rules that would make the disagreement even more dramatic. We, therefore, conclude that the symmetry-violating terms in the Hamiltonian must be induced by the transverse field.

Let us now write the total Hamiltonian as

$$\mathcal{H} = -DS_z^2 + g\mu_B H_z S_z - g\mu_B H_x S_x, \tag{7.8}$$

and demonstrate that the strengths of the transverse components of the dipole and hyperfine fields in $Mn_{12}Ac$ provide a quantitative explanation for the data, too. At $H_\parallel = 0$ the exact perturbation result for the tunneling splitting of each level m was found by Garanin [139]. Following his line of argument, one can generalize his formula to the nth resonance at $H_\parallel = nH_0$:

$$\Delta_m^{(n)} = \frac{2D}{[(2m - n - 1)!]^2}\left(\frac{(S + m - n)!(S + m)!}{(S - m)!(S - m + n)!}\right)^{1/2}\left(\frac{g\mu_B H_\perp}{2D}\right)^{2m-n}. \tag{7.9}$$

Divided by $\hbar$, it gives the tunneling rate from a particular excited level m tuned to the nth resonance. The answer is valid at small $H_\perp$. For $n = 0$ and $m = S$ it can also be obtained by the instanton method. (One can recognize the WKB exponent, $\exp[-2S\ln(H_a/H_\perp)]$, and the square-root tunneling prefactor.) For $S = 10$ one can also obtain all energy levels numerically by diagonalizing the Hamiltonian (7.8). As has been discussed at the beginning of this chapter, the average magnetic dipole field and the average hyperfine field experienced by a $Mn_{12}O_{12}$ molecule must be of the order of 100 G. Both fields have transverse components that, in the absence of an external transverse field, must be responsible for the transitions. The hyperfine and dipole fields also have random longitudinal components, but, just to illustrate our point, consider for a moment the case of zero bias, $H_\parallel = 0$. The tunneling rate from different levels at $n = 0$, computed from Eq. (7.9) at $H_\perp = 100$ G is presented in Table 7.1. One can see from Table 7.1 that the lifetime with respect to tunneling increases very fast, from nanoseconds to the lifetime of the universe and beyond, as one goes from the top to the bottom of the right-hand well in Fig. 7.20.

Table 7.1 Tunneling rates from various spin levels of the
Mn$_{12}$O$_{12}$ molecule at $H_\parallel = 0$ and $H_\perp = 100\,\mathrm{G}$ [130]

m	Tunneling rate (Hz)
1	3.2×10^8
2	1.1×10^5
3	3.5×10^0
4	2.3×10^{-5}
5	4.7×10^{-11}
6	3.7×10^{-17}
7	1.2×10^{-23}
8	1.7×10^{-30}
9	1.1×10^{-37}
10	2.1×10^{-45}

Many theoretical works on MQT in a double-well system have concerned tunneling between ground state levels at zero or very low temperature, for which excited levels are irrelevant. A well-known result is that tunneling in such a system freezes as soon as the asymmetry of the double well exceeds the tunneling splitting. The physical picture behind this statement is very simple. Owing to the tunneling, the ground state levels acquire a finite width Δ. As soon as the bias between the wells (in our case, produced by $H_\parallel$) exceeds Δ, the levels no longer overlap and the transitions between them freeze. In MQT problems the tunneling rate is exponentially small and so must be the bias. It is often very difficult or impossible to comply with that condition in experiments. For, e.g., a small ferromagnetic particle this may require an unphysically small external magnetic field. In this connection, it is important to emphasize that experimental constraints on observing tunneling, and even quantum coherence, between excited levels can be somewhat less restrictive. Indeed, as one moves up from the ground-state level to the levels at the top of the well, the tunneling splitting grows dramatically (see Table 7.1), so that the requirement that the system be in resonance can be relaxed. Insofar as the MQC is concerned, there is an additional constraint [45], however, namely that the splitting must be greater than the interaction with the environment. For the excited levels the measure of this interaction is the width of the level with respect to the decay down the well, the value of which should be of the order of the pre-exponential factor in the transition rate. For Mn$_{12}$Ac, $\nu \simeq 10^7\,\mathrm{s}^{-1}$, and upon applying a transverse field, one can progressively increase the depth of the level for which the tunneling splitting is greater than the dissipation width. The real picture may be somewhat more complicated because of the fine structure of m-levels due to the hyperfine interaction. The latter splits each m-level into hundreds of sublevels, making it more appropriate to talk about an m-band of width about 10^9 s^{-1} (about 100 G).

The total relaxation rate in Mn$_{12}$Ac must be a sum over all processes which rotate the spin of a Mn$_{12}$O$_{12}$ molecule from one well to another. Alongside thermal overbarrier transitions, these processes include tunneling from excited levels. Which levels contribute most depends on their thermal population and the tunneling rate. In practice, the effect of the resonant tunneling from excited levels can be observed only within a limited temperature range below T_B. In Mn$_{12}$Ac, very fortunately for the experimentalist, T_B falls within the kelvin range. This is due to the very high anisotropy of this material, $H_a = 9.2$ T. Above T_B transitions are dominated by thermal effects, whereas well below T_B they become completely blocked since the upper levels, for which tunneling is significant, become depopulated. This can be corrected by decreasing the barrier with the help of the longitudinal field, but, nonetheless, to observe resonant tunneling, the temperature must be below T_B. This is the reason why one has to go to lower temperatures to observe resonances corresponding to large n. It is easy to estimate that, e.g., resonance number 19 at $H_\parallel = 8.75$ T must become apparent at around 10 T.

Both the rate of thermal activation to a particular level, $\Gamma_m \simeq \nu \exp[-(E_m - E_{-10})/T]$ and the tunneling rate from that level, Eq. (7.9), depend exponentially on m. Consequently, for any particular m the tunneling rate, unless by accident, will be either very large or very small compared with the thermal rate. The fact that it is very large would mean that the spin goes across the barrier as soon as it reaches that level, which is equivalent to no barrier for the corresponding m. The first level from the top for which the tunneling rate becomes small compared with the thermal activation rate can be called the *blocking level*, m_B. It determines the effective height of the barrier, $U_{eff} = E_{m_B} - E_{-10}$, at resonance. Apart from that, the relaxation rate at resonance must obey the Arrhénius law, for it becomes equal to the rate of thermal activation to the blocking level. For relaxation at resonance, the value of m_B depends both on $H_\parallel$ and on $H_\perp$. At $H_\parallel = 0$, one obtains from the above formulas that $m_B = -3$ at $H_\perp = 100$ G and progressively decreases to $m_B = -10$ as $H_\perp$ goes from 100 G to H_a. The dependence of m_B on $H_\perp$ is roughly in accordance with the value of the effective barrier at resonance obtained from the measurements of T_B and from relaxation measurements [134,135]. Note, however, that the average dependence of U_{eff} on $H_\perp$ through the dependence $m_B(H_\perp)$ is equivalent to the classical formula, $H_{eff} = H_a(1 - H_\perp/H_a)^2$. Therefore, a more subtle analysis of the effect of the transverse field is needed in order to extract features that are specific to the quantized spin levels. One such feature should be a step-wise, rather than smooth, dependence of U_{eff} on the transverse field, reflecting the discrete nature of the variable m_B in the dependence $U_{eff}(m_B)$. Away from the resonance, the relaxation must be purely thermal and determined by the full barrier, U. This is roughly in accordance with observations.

A comprehensive theory of the transition rate in Mn$_{12}$Ac has been developed on the basis of the density matrix formalism [140]. The theory takes into account interactions of spin 10 with phonons and nuclear spins, as well as with magnetic dipole fields created by other Mn$_{12}$O$_{12}$ molecules. It is too involved to enter this book in detail. Here we shall only emphasize one prediction of the theory, which may apply to many different magnetic systems. It turns out [141] that the model described by the Hamiltonian given by Eq. (7.8) has both first- and second-order transitions from the thermal to the quantum behavior in a sense explained in section 7.2. This was first observed in perturbation theory [140] by plotting m_B versus T for Mn$_{12}$Ac, see Fig. 7.21. At $h_x = H_x/H_{an} = 0.1$, there is a discontinuity in the dependence of m_B on T for all S, in the sense that m_B, at a certain temperature $T_c^{(1)}$, changes by more than 1. This discontinuity corresponds to the first-order crossover from the quantum to the thermal regime. At $h_x = 0.2$, however, m_B changes continuously with increasing T, which corrsponds to the second-order crossover. Thus, both kinds of crossover are possible, depending on the strength of the transverse field. These findings have been confirmed [141] by mapping the spin Hamiltonian onto the particle Hamiltonian according to Eq. (3.8). At $H_z = 0$, the equivalent potential for a particle that results in the same low-lying energy levels is

$$U = \left(S + \tfrac{1}{2}\right)^2 D\left(h_x^2 \sinh^2 x - 2h_x \cosh x\right). \tag{7.10}$$

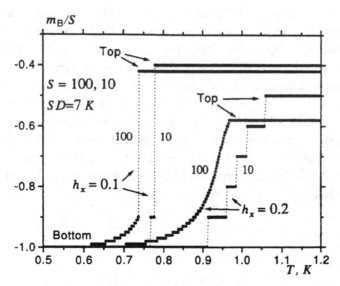

Figure 7.21   The temperature dependence of the group of levels m_{TAT} making the dominant contribution into the thermally assisted tunneling. (Cited from [140].)

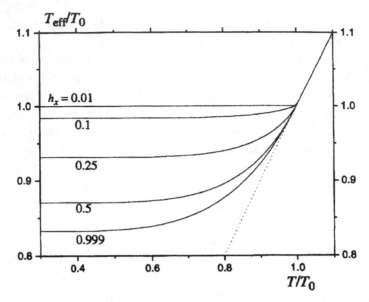

Figure 7.22   The dependences of the escape temperature (here refered as T_{esc}) on T for the different values of the transverse field. (Cited from [141].)

According to section 2.7, the second- or first-order transition from the quantum to the thermal regime occurs depending on whether the period of oscillations in the inverted potential is a monotonic or a non-monotonic function of energy. It is easy to show [141] that, in the limit of large spin, $H_x < H_{an}/4$ results in a first-order transition, whereas $H_x > H_{an}/4$ yields a second-order transition. The corresponding dependence of T_{esc} on T is shown in Fig. 7.22. Note that thermal and quantum corrections to the escape rate smear the transition in the vicinity of the crossover temperature. Nevertheless, for large S the differences between the first- and the second-order crossover must be apparent.

We would like to conclude this chapter by expressing the opinion that the study of thermally assisted tunneling in magnetic molecules opens a fascinating chapter in the field of MQT and MQC both for theory and for experiment. The main advantage of these systems from the experimental point of view is the precise knowledge and good control of the parameters. From a theoretical point of view, they pose the challenging problem of MQT between excited levels of a double-well system.

References

[1] L. D. Landau and E. M. Lifshitz, *Sov. Phys. JETP* **8**, 153 (1935).

[2] E. M. Chudnovsky, *Sov. Phys. JETP* **50**, 1035 (1979).

[3] J. S. Langer, *Ann. Phys.* **41**, 108 (1967); A. M. Polyakov, *Nucl. Phys. B* **120**, 429 (1977).

[4] E. M. Chudnovsky and L. Gunther, *Phys. Rev. Lett.* **60**, 661 (1988).

[5] E. M. Chudnovsky and L. Gunther, *Phys. Rev. B* **37**, 9455 (1988).

[6] B. Barbara and E. M. Chudnovsky, *Phys. Lett. A* **145**, 205 (1990).

[7] A. O. Caldeira and A. J. Leggett, *Phys. Rev. Lett.* **46**, 211 (1981); *Ann. Phys. (N.Y.)* **149**, 374 (1983).

[8] M. Enz and R. Schilling, *J. Phys. C* **19**, 1765 (1986).

[9] J. L. van Hemmen and A. Sütő, *Physica B* **141**, 37 (1986).

[10] Anupam Garg, *Phys. Rev. Lett.* **70**, 1541 (1993).

[11] N. V. Prokof'ev and P. C. E. Stamp, *J. Phys.: Condens. Matter* **5**, L663 (1993).

[12] J. Tejada and X. X. Zhang, *J. Magn. Magn. Mater.* **140–144**, 1815 (1995).

[13] D. D. Awschalom, J. F. Smyth, G. Grinstein, D. T. DiVincenzo, and D. Loss, *Phys. Rev. Lett.* **68**, 3092 (1992).

[14] W. Wernsdorfer, B. Doudin, D. Mailly, K. Hasselbach, A. Benoit, J. Meier, J.-Ph. Ansenermet, and B. Barbara, *Phys. Rev. Lett.* **77**, 1873 (1996); W. Wernsdorfer, E. Bonet Orozco, K. Hasselbach, A Benoit, D. Maily, O. Kubo, H. Nakano, and B. Barbara, *Phys. Rev. Lett.* **79**, 4014 (1997).

[15] K. Hong and N. Giordano, in *Quantum Tunneling of Magnetization – QTM '94*, ed. by L. Gunther and B. Barbara, John Wiley & Sons, New York (1995).

[16] J. R. Friedman, M. P. Sarachik, J. Tejada, and R. Ziolo, *Phys. Rev. Lett.* **76**, 3830 (1996).

[17] D. P. DiVincenzo, *Science* **270**, 255 (1995).

[18] H. A. Kramers, *Physica (Utrecht)* **7**, 284 (1940).

[19] I. Affleck, *Phys. Rev. Lett.* **46**, 388 (1981); A. I. Larkin and Yu. N. Ouchinnikov, *Zh. Eksp. Teor. Fiz.* **38**, 111 (1983) [*JETP Lett.* **37**, 382 (1983)].

[20] R. P. Feynman and A. R. Hibbs, *Quantum Mechanics and Path Integrals*, McGraw-Hill, New York (1965).

[21] A. M. Polyakov, *Gauge Fields and Strings*, Harwood Academic Publishers, London (1987).

[22] S. Coleman, *Phys. Rev. D* **15**, 2929 (1977); C. C. Callan and S. Coleman, *Phys. Rev. D* **16**, 1762 (1977).

[23] E. M. Chudnovsky, *Phys. Rev. A* **46**, 8011 (1992).

[24] H. Grabert and U. Weiss, *Phys Rev. Lett.* **53**, 1787 (1984); A. I. Larkin and Yu. Ouchinnikov, *Zh. Eksp. Teor. Fiz.* **86**, 719 (1984) [*Sov. Phys. JETP* **59**, 420 (1984)]; W. Zwerger, *Phys. Rev. A* **31**, 1745 (1985); P. S. Riseborough, P. Hänggi, and E. Freidkin, *Phys. Rev. A* **32**, 489 (1985).

[25] A. J. Leggett, *Phys. Rev. B* **30**, 1208 (1984).

[26] E. H. Frei, S. Shtrikman, and D. Treves, *Phys. Rev.* **106**, 446 (1957); W. F. Brown, *Phys. Rev.* **105**, 1479 (1957); W. F. Brown, *J. Appl. Phys.* **39**, 993 (1968).

[27] P. A. Serena and N. García, in *Proceedings of the NATO Advanced Research Workshop on Quantum Tunneling of Magnetization*, Kluwer, Dordrecht (1995).

[28] R. H. Kodama, A. E. Berkowitz, E. J. McNiff Jr, and S. Foner, *Phys. Rev. Lett.* **77**, 394 (1996).

[29] E. M. Chudnovsky, *Phys. Rev. Lett.* **72**, 3433 (1994).

[30] A. Aharoni and S. Shtrikman, *Phys. Rev.* **109**, 1522 (1958); A. Aharoni, *Phys. Status Solidi* **16**, 3 (1966).

[31] D. Stauffer, *Solid State Commun.* **18**, 533 (1976).

[32] E. Freidkin, *Field Theories of Condensed Matter*, Addison-Wesley, New York (1991).

[33] D. Loss, D. P. DiVincenzo, and G. Grinstein, *Phys. Rev. Lett.* **69**, 3232 (1992).

[34] J. von Delft and C. L. Henly, *Phys. Rev. Lett.* **69**, 3236 (1992).

[35] E. M. Chudnovsky and D. P. DiVincenzo, *Phys. Rev. B* **48**, 10 548 (1993).

[36] O. B. Zaslavskii, *Phys. Rev. B* **42**, 992 (1990).

[37] M. C. Miguel and E. M. Chudnovsky, *Phys. Rev. B* **54**, 388 (1996).

[38] G. Sharf, *Ann. Phys. (N.Y.)* **83**, 71 (1974); O. B. Zaslavskii and V. V. Ulyanov, *Zh. Eksp. Teor. Fiz.* **87**, 1724 (1984) [*Sov. Phys. JETP* **60**, 991 (1994)].

[39] E. M. Chudnovsky, in *Proceedings of the NATO Advanced Research Workshop on Quantum Tunneling of Magnetization*, Kluwer, Dordrecht (1995).

[40] I. V. Krive and O. B. Zaslavskii, *J. Phys.: Condens. Matter* **2**, 9457 (1990).

[41] Anupam Garg and G.-H. Kim, *Phys. Rev. Lett.* **63**, 2512 (1989); *Phys. Rev.* **43**, 712 (1991).

[42] E. M. Chudnovsky, O. Iglesias, and P. C. E. Stamp, *Phys. Rev. B* **46**, 5392 (1992).

[43] G. Tatara and H. Fukuyama, *Phys. Rev. Lett.* **72**, 772 (1994).

[44] E. M. Chudnovsky, *Phys. Rev. Lett.* **72**, 1134 (1994).

[45] A. J. Leggett, S. Chakravarty, A. T. Dorsey, P. A. Fisher, Anupam Garg, and W. Zwerger, *Rev. Mod. Phys.* **59**, 1 (1987).

[46] E. Merzbacher, *Quantum Mechanics*, John Wiley & Sons, New York (1970).

[47] A. Ferrera and E. M. Chudnovsky, *Phys. Rev. B* **53**, 354 (1996).

[48] T. Egami, *Phys. Status Solidi A* **20**, 157 (1973); *Phys. Status Solidi B* **57**, 211 (1973).

[49] B. Barbara, G. Fillion, D. Gignoux, and R. Lemaire, *Solid State Commun.* **10**, 1149 (1972).

[50] O. Bostanjoglo and H. P. Gemund, *Phys. Status Solidi A* **17**, 115 (1973); **48**, 481 (1978).

[51] A. P. Malozemoff and J. C. Slonczewski, *Magnetic Domain Walls in Bubble Materials*, Academic, New York (1979).

[52] J. Clarke, A. N. Cleland, M. H. Devoret, D. Esteve, and J. M. Martinis, *Science* **239**, 992 (1988).

[53] P. C. E. Stamp, *Phys. Rev. Lett.* **66**, 2802 (1991).

[54] V. G. Baryakhtar, B. A. Ivanov, and M. V. Chetkin, *Usp. Fiz. Nauk* **146**, 417 (1985) [*Sov. Phys. Usp.* **28**, 563 (1985)].

[55] E. M. Chudnovsky, *J. Magn. Magn. Mater.* **140–144**, 1821 (1995).

[56] Y. B. Kim and M. J. Stephen in *Superconductivity*, ed. by R. D. Park, Marcel Dekker, New York (1969), chapter 19.

[57] J. Tejada, E. M. Chudnovsky, and A. García, *Phys. Rev. B* **47**, 11 552 (1993).

[58] N. B. Kopnin and V. E. Kravtsov, *Zh. Eksp. Teor. Fiz. – Pis'ma* **23**, 631 (1976) [*JETP Lett.* **23**, 578 (1976)]; *Zh. Eksp. Teor. Fiz.* **71**, 1644 (1976) [*Sov. Phys. JETP* **44**, 861 (1976)].

[59] J. M. Harris, Y. F. Yau, O. K. C. Tsui, Y. Matsuda, and N. P. Ong, *Phys. Rev. Lett.* **73**, 1711 (1994).

[60] E. M. Chudnovsky, A. Ferrera, and A. Vilenkin, *Phys. Rev. B* **51**, 1181 (1985).

[61] P. W. Anderson and Y. B. Kim, *Rev. Mod. Phys.* **36**, 39 (1964).

[62] See for a review G. Blatter, M. V. Feigel'man, V. B. Geshkenbein, A. I. Larkin, and V. M. Vinokur, *Rev. Mod. Phys.* **66**, 1125 (1994).

[63] J. M. Hernandez, X. X. Zhang, and J. Tejada, *J. Appl. Phys.* **79**, 4686 (1996).

[64] J. Tejada, X. X. Zhang, and J. M. Hernandez, *Magnetic Hysteresis Phenomena in Novel Materials*, NATO ASI Conference at Mykonos, Greece, ed. by G. C. Hadjipanayis, Kluwer, Dordrecht (1996).

[65] J. González-Miranda and J. Tejada, *Phys. Rev. B* **49**, 3867 (1994).

[66] C. Sánchez, J. González-Miranda, and J. Tejada (unpublished).

[67] E. Vincent, J. Hamman, P. Prene, and E. Tronc, *J. Physique* **4**, 273 (1994).

[68] X. X. Zhang, R. F. Ziolo, E. C. Kroll, X. Bohigas, and J. Tejada, *J. Magn. Magn. Mater.* **140–145**, 1853 (1995).

[69] R. Sappey, E. Vincent, J. Hamman, F. Chaput, J. Boilot, and D. Zins, *Magnetic Hysteresis Phenomena in Novel Materials*, NATO ASI Conference at Mykonos, Greece, ed. by G. C. Hadjipanayis, Kluwer, Dordrecht (1996).

[70] J. Tejada, R. F. Ziolo, and X. X. Zhang, *Chem. Mater.* **8**, 1784 (1996) (review).

[71] S. Chikazumi and S. H. Charap, *Physics of Magnetism,* John Wiley & Sons, New York (1964).

[72] Anupam Garg, *Phys. Rev. Lett.* **71**, 4249 (1993); *Ibid.* **74**, 1458 (1995).

[73] W. Krakow and B. M. Siegel, *J. Appl. Crystallogr.* **9**, 325 (1976).

[74] P. M. Harrison, F. A. Fischbach, T. G. Hoy, and G. H. Haggis, *Nature* **216**, 1188 (1967).

[75] G. H. Haggis, *J. Mol. Biol.*, **14**, 598 (1965).

[76] S. H. Bell, M. P. Weir, D. P. E. Dickson, J. F. Gibson, G. A. Sharp, and T. J. Peters, *Biochim. Biophys. Acta* **787**, 227 (1994).

[77] T. G. St. Pierre, D. H. Jones, and D. P. E. Dickson, *J. Magn. Magn. Mater.*, **69**, 276 (1987).

[78] R. Bauminger and I. Nowik, *Hyperfine Interactions*, **50**, 484 (1989).

[79] S. Gider, D. D. Awschalom, T. Douglas, S. Mann, and M. Chaparala, *Science* **268**, 77 (1995).

[80] J. Tejada and X. X. Zhang, *J. Phys.: Condens. Matter* **6**, 263 (1994).

[81] S. H. Kilcoyne and R. Cywinski, *J. Magn. Magn. Mater.* **140–144**, 1466 (1995).

[82] C. Paulsen, L. C. Sampaio, B. Barbara, R. Tucoulou-Tachoueres, D. Fruchart, A. Marchand, J. L. Tholence, and M. Uehara, *Europhys. Lett.* **19**, 643 (1992).

[83] R. H. Kodama, C. L. Seamon, A. E. Berkowitz, and M. B. Maple, *J. Appl. Phys.* **75**, 5639 (1994).

[84] M. N. M. Ibrahim, S. Darwish, and M. M. Seehra, *Phys. Rev. B* **51**, 2955 (1995).

[85] M. J. O'Shea and P. Perera, *J. Appl. Phys.*, **76**, 6174 (1994); P. Perera and M. J. O'Shea, *Phys. Rev. B* **53**, 3381 (1996).

[86] X. X. Zhang, Ll. Balcells, J. M. Ruiz, O. Iglesias, J. Tejada, and B. Barbara, *Phys. Lett. A* **163**, 130 (1992).

[87] J. Tejada, X. X. Zhang, and E. M. Chudnovsky, *Phys. Rev. B* **47**, 14977 (1993).

[88] J. I. Arnaudas, A. del Moral, C. de la Fuente, M. Ciria, and P. A. J. de Groot, *Phys. Rev. B* **50**, 547 (1994).

[89] E. M. Chudnovsky, *Phys. Rev. B* **47**, 9102 (1993).

[90] B. Barbara and M. Uehara, *Physica* **86–88**, 1481 (1977).

[91] M. Uehara, B. Barbara, B. Dieny, and P. C. E. Stamp, *Phys. Lett. A* **114**, 23 (1986).

[92] M. Uehara and B. Barbara, *J. Physique* **47**, 235 (1986).

[93] J. E. Bourée and J. Hammann, *J. Physique* **36**, 391 (1975).

[94] J. D. Gordon, G. Gorodetzky, and R. M. Hornreich, *J. Magn. Magn. Mater.* **3**, 288 (1976).

[95] E. F. Bertaut, J. Chappert, J. Mareschal, J.P. Rebouillet, and J. Sivardière, *Solid State Commun.* **5**, 293 (1967).

[96] M. Belahovsky, J. Chappert, T. Rouskov, and J. Sivardière, *J. Physique* **32**, C1-482 (1971).

[97] J. Tejada, X. X. Zhang, A. Roig, O. Nikolov, and E. Molins, *Europhys. Lett.* **30**, 227 (1995).

[98] X. X. Zhang, J. M. Hernandez, J. Tejada, R. Solé, and X. Ruiz, *Phys. Rev. B* **53**, 3336 (1996).

[99] Y. Yeshurun, A. P. Malozemoff, and A. Shaulov, *Rev. Mod. Phys.* **69**, 911 (1996).

[100] A. C. Mota, A. Pollini, P. Visani, K. A Müller, and J. G. Bednorz, *Phys. Rev. B* **36**, 4011 (1988).

[101] D. Prost, L. Fruchter, I. A. Campbell, N. Motohira, and M. Koncyzkowski, *Phys. Rev. B* **47**, 3457 (1993).

[102] A. Hamzic, L. Fruchter, and I. A. Campbell, *Nature* **345**, 515 (1990).

[103] L. Fruchter, A. P. Malozemoff, I. A. Campbell, J. Sanchez, M. Konczykowski, R. Griessen, and F. Holtzberg, *Phys. Rev. B* **43**, 8709 (1991).

[104] A. C. Mota, P. Visani, A. Pollini, G. Juri, and D. Jérome, *Physica C* **153–155**, 1153 (1988).

[105] A. C. Mota, P. Visani, and A. Pollini, *Physica C* **153–155**, 441 (1988).

[106] A. V. Mitin, *Sov. Phys. JETP* **66**, 335 (1987).

[107] X. X. Zhang, A. García, J. Tejada, Y. Xin, and K. W. Wong, *Physica C* **232**, 99 (1994).

[108] A. García, X. X. Zhang, J. Tejada, M. Manzel, and H. Bruchlos, *Phys. Rev. B* **50**, 9439 (1995).

[109] G. Blatter, V. B. Geshkenbein, and V. M. Vinokur, *Phys. Rev. Lett.* **66**, 3297 (1991).

[110] X. X. Zhang, J. M. Hernandez, J. Tejada, and R. F. Ziolo, *Phys. Rev. B* **64**, 4101 (1996)

[111] E. Krotenko, X. X. Zhang, and J. Tejada, *J. Magn. Magn. Mater.* **150**, 119 (1995).

[112] S. Vitale, M. Cerdonio, G. A. Prodi, A. Cavalleri, P. Faltari, and A. Maraner, in *Quantum Tunneling of Magnetization – QTM '94*, ed. by L. Gunther and B. Barbara, John Wiley & Sons, New York (1995).

[113] F. Luis, Ph.D. Dissertation. Universidad de Zaragoza (1997).

[114] F. Luis, J. Bartolomé, J. Tejada, and E. Martinez, *J. Magn. Magn. Mater.* **157–158**, 266 (1996).

[115] A. Maraner, X. Xhang, A. Cavalleri, J. Tejada, and S. Vitale, *J. Appl. Phys.* **79**, 5406 (1996).

[116] F. Luis, E. del Barco, X. X. Zhang, J. M. Hernandez, J. Bartolomé, and J. Tejada (unpublished).

[117] H. Kobayashi, N. Hatano, A. Terai, and M. Suzuki, *Physica A* **226**, 137 (1996).

[118] J. L. Dorman and D. Fiorani (editors), *Magnetic Properties of Fine Particles,* North Holland, Amsterdam (1992).

[119] J. Tejada, X. X. Zhang, E. del Barco, J. M. Hernandez, and E. M. Chudnovsky, *Phys. Rev. Lett.* **79**, 1754 (1997).

[120] S. Gider, D. D. Awschalom, T. Douglas, K. Wong, S. Mann, and G. Cain, *J. Appl. Phys.* **79**, 5324 (1996).

[121] J. F. Friedman, U. Voskoboynik and M. P. Sarachick, *Phys. Rev. B* **56**, 10793 (1997).

[122] R. Sappey, E. Vincent, H. Hadacek, F. Chaput, J.P. Boilot, and D. Zius, *Phys. Rev. B* **56**, 14551 (1997).

[123] J. Tejada, *Science* **272**, 424 (1996); Anupam Garg, *Ibid.* **272**, 424 (1996); S. Gider, D. D. Awschalom, D. P. DiVincenzo, and D. Loss, *Ibid.* **272**, 425 (1996).

[124] E. C. Stoner and E. P. Wohlfarth, *Philos. Trans. London Series A* **240**, 599 (1948).

[125] M. Lederman, S. Schultz, and M. Ozaki, *Phys. Rev. Lett.* **73**, 1986 (1994).

[126] W. Wernsdorfer, E. B. Orozco, K. Hasselbach, A. Benoit, B. Barbara, N. Demoncy, A. Loiseau, H. Pascard, and D. Mailly, *Phys. Rev. Lett.* **78**, 1791 (1997).

[127] W. Wernsdorfer, K. Kasselbach, P. Mailly, B. Barbara, A. Benoit, L. Thomas, and G. Suran in *Quantum Tunneling of Magnetization – QTM '94*, ed. by L. Gunther and B. Barbara, John Wiley & Sons, New York, p. 227 (1995).

[128] T. Lis, *Acta Crystallogr. B* **36**, 2042 (1980).

[129] R. Sessoli, H. L. Tsai, A. R. Schake, S. Wang, J. B. Vincent, K. Folting, D. Gatteschi, G. Christou, and D. N. Hendrikson, *J. Am. Chem. Soc.* **115**, 1804 (1993).

[130] A. Caneschi, D. Gatteschi, R. Sessoli, A. L. Barra, L. C. Brunel, and M. Guillot, *J. Am. Chem. Soc.*, **113**, 5873 (1991).

[131] M. A. Novak and R. Sessoli in *Quantum Tunneling of Magnetization – QTM '94*, ed. by L. Gunther and B. Barbara, John Wiley & Sons, New York, p. 171 (1995).

[132] C. Paulsen and J. G. Park in *Quantum Tunneling of Magnetization – QTM '94*, ed. by L. Gunther and B. Barbara, John Wiley & Sons, New York, p. 193 (1995).

[133] F. Hartmann-Boutron, P. Politi, and J. Villain *Int. J. Mod. Phys.* **10**, 2577 (1996).

[134] J. M. Hernandez, X. X. Zhang, F. Luis, J. Tejada, J. R. Friedman, M. P. Sarachik, and R. Ziolo, *Phys. Rev. B* **55** 5858 (1997).

[135] J. M. Hernandez, X. X. Zhang, F. Luis, J. Bartolomé, J. Tejada, and R. Ziolo, *Europhys. Lett.* **35**, 301 (1996).

[136] L. Thomas, F. Lionti, R. Ballou, D. Gatteschi, R. Sessoli, and B. Barbara, *Nature* **383**, 145 (1996).

[137] J. Tejada, X. X. Zhang, J. M. Hernandez, J. R. Friedman, M. P. Sarachik and R. F. Ziolo in *Magnetic Hysteresis Phenomena in Novel Materials*, NATO ASI Conference at Mykonos, Greece, ed. by G. C. Hadjipanayis, Kluwer, Dordrecht (1996).

[138] F. Luis, J. Bartolomé, J. F. Fernández, J. Tejada, J. M. Hernandez, X. X. Zhang, and R. Ziolo, *Phys. Rev. B* **55**, 11 448 (1997).

[139] D. A. Garanin, *J. Phys. A* **24**, L61 (1991).

[140] D. A. Garanin and E. M. Chudnovsky *Phys. Rev. B* **56**, 11102 (1997).

[141] E. M. Chudnovsky and D. A. Garanin *Phys. Rev. Lett.* **79**, 4469 (1997).

Index